SPRINGER
LAB MANUALS

W0232170

Springer-Verlag Berlin Heidelberg GmbH

R. CURTIS BIRD     BRUCE F. SMITH (EDS.)

# Genetic Library Construction and Screening

## Advanced Techniques and Applications

With 30 Figures

 Springer

Dr. R. Curtis Bird

Department of Pathobiology
Auburn University
Auburn, AL 36849-5519
USA

e-mail: birdric@betmed.auburn.edu

Dr. Bruce F. Smith

Scott-Ritchey Research Center
Auburn University
Auburn, AL 36849
USA

e-mail: smithbf@auburn.edu

ISBN 978-3-642-47733-1

Applied for Library of Congress Cataloging-in-Publication Data

Die Deutsche Bibliothek – CIP-Einheitsaufnahme

Genetic library construction and screening : advanced techniques and
applications / R. Curtis Bird ; Bruce F. Smith (ed.). – Berlin ; Heidelberg
; New York ; Barcelona ; Hong Kong ; London ; Milan ; Paris ; Tokyo :
Springer, 2002
  (Springer lab manuals)
  ISBN 978-3-642-47733-1      ISBN 978-3-642-56408-6 (eBook)
  DOI 10.1007/978-3-642-56408-6

http://www.springer.de

© Springer-Verlag Berlin Heidelberg 2002
Originally published by Springer-Verlag Berlin Heidelberg New York in 2002

Cover design: *design & production* GmbH, D-69121 Heidelberg
Typesetting: Kröner, D-69115 Heidelberg

SPIN 10531590    39/3130 YK – 5  4  3  2  1  0 – Printed on acid free paper

# Preface

In recent years, it has seemed as though classical cloning strategies were taking something of a back seat to new advances in the development of applied polymerase chain reaction strategies. The tried and established methods of library construction and recombinant DNA cloning pioneered during the 1970s and early 1980s provided many of the bedrock approaches used to find cDNA and genomic clones of significance and made possible accessible DNA sequencing technologies that did not require exotic or even hazardous chemistry. Today, these technologies are enjoying a renaissance as they find new utility in novel guises combined with the newer approaches and technologies based on polymerase chain reaction. Although it is unlikely that any scientist today would employ all of the technologies presented in this volume, every scientist working in the disciplines of molecular biology and genetics should be familiar with all of them and will likely need a working knowledge or have an application for some of them. This volume is designed to provide an easily accessible introduction to what are rapidly becoming core strategies in the application of modern recombinant DNA technologies. These approaches should be of value to both experienced scientists seeking to apply these technologies to their systems as well as to scientists and graduate students, in the more formative stages of development as molecular biologists, wishing to apply these powerful technologies.

The first section of this volume, encompassing the first five chapters, addresses directly the applications of polymerase chain reaction to a variety of problems in DNA cloning that are, or have been, extremely challenging using more traditional approaches and technologies. In truth, what we used to caution in our laboratories, were cloning experiments only to be undertaken by scientists and students experienced in advanced cloning technology are experiments now well within the grasp of those at a more formative level of training and experience. Although still not for the novice, these technologies are well within the abilities of graduate students and scientists new to the discipline once a basic competence in small volume handling, RNase-free techniques and the theoretical precepts of recombinant DNA have been mastered. These chapters introduce students and scientists to exciting applications of polymerase chain reaction that render manageable applications such as simple cDNA cloning and transcript mapping, mutagenesis and even the challenging cloning of very long transcripts. Additionally, this section demonstrates how to clone libraries successfully from limiting amounts of total RNA, an application that was considered extremely difficult using earlier approaches.

The second section of this volume, encompassing the sixth and seventh chapters, describes modern approaches to subtractive cloning technologies that seek to select only those sequences of interest prior to the cloning step. When originally attempted in the 1980s, these pre-screening approaches to target sequence enrichment proved extremely challenging and even beyond the abilities of some given the limitations of the technologies available. Now subtractive approaches, with or without applications of polymerase chain reaction, can be successfully undertaken by scientists of many levels of experience in the application of recombinant DNA technology. These improvements have provided a major advance in our ability to detect and clone important and relatively low abundance sequences.

In the last section, composed of chapters eight through eleven, a variety of specialized expression cloning and library screening strategies are introduced. These begin with two variations on two-hybrid analysis that provide, first, a strategy for investigation of protein-protein interactions while the second provides a strategy for investigation of protein-nucleic acid interactions. Both provide the means to probe extremely subtle interactions not amenable to precipitation or chromatographic approaches requiring much more stable macromolecular interactions. The possibilities for investigation of the subtle touch-and-go types of macromolecular interactions often encountered between signaling proteins or between transcription factors and their DNA bind sites are now greatly enhanced, thus expanding the possibilities for gene discovery in this discipline.

The last two chapters describe the application of alternative strategies to common experimental problems. The first is to provide high throughput screening for applications where functional screening is not possible yet, where many thousands of clones must be interrogated in an informative way to identify needed clones. This approach is particularly useful as it employs fluorescent non-radioactive imaging technologies compatible with many of the modern automated approaches to high throughput handling of recombinant clones. Lastly, the final chapter provides insight into the realm of alternative approaches to creating artificial affinity reagents - artificial antibodies if you will - that can provide what is essentially an infinite array of peptides from which to select ligand-binding complexes. Additionally, because these complexes are in reality bacteriophages, they come with many thousands of sites for secondary binding of directly or indirectly fluorescently labeled antibodies. This capacity has the potential to drive detection limits to ranges undreamt of for antibodies and to provide affinity reagents without the need to consider antigenicity.

The potential of these reagents in the realm of gene discovery, diagnostics as well as therapeutics is nothing short of breathtaking when the many applications being investigated are considered. However, perhaps the most exciting applications of all come from the investigation of these technologies and the development and exploration of prophylactic treatments for important diseases and threats. In this realm, these technologies may soon demonstrate their most important value in the campaign to improve human and animal well-being and to alleviate suffering due to the ravages of disease. This is particularly true in an era where, in addition to the natural development of disease, humans can threaten to foment new biological plagues.

While we have been absorbed in reading, writing and editing this exciting project, there are many people that we would like to acknowledge and thank for their special contributions. In the development of a volume such as this, their contributions are of the highest import because without their input and encouragement this project would not have been possible. Though no list can ever be inclusive and complete, the individual authors of each chapter and our very patient and supportive editor in the life sciences editorial office of Springer-Verlag in Heidelberg, Dr. Jutta Lindenborn, deserve the most credit. Their contributions and the quality of the text speak for themselves. Additionally, we would like to thank our wives/collaborators (they are both), our families, and those in our laboratories for their continuing efforts, support, and encouragement. Their contributions, while less tangible within the pages of this volume, are no less important in the effects they have on the quality of our efforts. Additionally, we would like to thank the College of Veterinary Medicine at Auburn University and our faculty colleagues for continued support, both financial and intellectual, during the completion of this project and for the rich scholarly environment in which we have spent so many years. Our gratitude to them all cannot be adequately expressed as each contributes his/her special flavor to an exotic, intellectually challenging and stimulatingly diverse working environment. Our gratitude is extended to them all.

Auburn University                                R. Curtis Bird, Ph.D.
December 2001                      Bruce F. Smith, V.M.D., Ph.D.

# Contents

# Contributors

ELLEN N. BEHREND
Department of Clinical Sciences
Auburn University
Auburn, AL 36849
USA
Phone: (334) 844-4690
Fax: (334) 844-6034
E-mail: behreen@auburn.edu

R. CURTIS BIRD
Department of Pathobiology
Auburn University
Auburn, AL 36849-5519
USA
Phone: (334)-844-4539
Fax: (334)-844-2652
E-mall: birdric~auburn.edu

JOHN O'BRIEN
Max-Planck-lnstitut
für Molekulare Genetik
Ihnestrasse 73
14195 Berlin
Germany

SHILPA J. BUCH
Marion Merrell Dow Laboratory
of Viral Pathogenesis
5000 East Wahl
University of Kansas
Medical Center
3901 Rainbow Boulevard
Kansas City, Kansas 66160-7240
USA

ALEX CHENCHIK
Gene Cloning and Analysis Group
Clontech Laboratories, Inc.
1020 East Meadow Circle
Palo Alto, CA 94303, USA
Phone: (650)-424-8222
Fax: (650)-354-0776
E-mail: yyzhu@clontech.com

HUGUES ROEST CROLLIUS
Genoscope
Centre National de Sequencage
2, rue Gaston Cremieux
91057 Evry Cedex, France
Phone: +33 (0) 1 60 87 25 64
Fax: +33 (0) 1 60 87 25 89
E-mall: hrc@genoscope.cns.fr

JOHN W. GOW
Department of Neurology
Institute of Neurological Sciences
Southern General Hospital,
University of Glasgow
1345 Govan Road
Glascow, G51 4TF, Scotland
Phone: 44 (0) 141 201 2465
Fax: 44 (0) 141 201 2515
E-mail: gora20@udcf.gla.ac.uk

FLORENCE Y. HSIEH
Gene Cloning and Analysis Department, BD Biosciences Clontech,
1020 East Meadow Circle, Palo Alto,
California 94303-4230, USA

BETTY C.B. HUANG
Rigel, Inc.
240 East Grand Aveneue,
South San Francisco,
CA 94080, USA
E-mail: bcbhuang@rigel.com

WEIWEN JIANG
c/o Bruce F. Smith
Scott-Ritchey Research Center
Auburn University
Auburn, AL 36849
USA
Phone: (334)-844-5951
Fax: (334)-844-5850
E-mail: smithbf@auburn.edu

ROBERT KEMPPAINEN
Department of Anatomy,
Physiology and Pharmacology
Auburn University
Auburn, AL 36849
USA
Phone: (334) 844-4425
Fax: (334) 844-5388
E-mail: kempprj@auburn.edu

HANS LEHRACH
Max-Planck-lnstitut
für Molekulare Genetik
Ihnestrasse 73
14195 Berlin
Germany

ROGER LI
Gene Cloning and Analysis Group
Clontech Laboratories, Inc.
1020 East Meadow Circie
Palo Alto, CA 94303
USA
Phone: (650)-424-8222
Fax: (650)-354-0776
E-mall: yyzhu@ciontech.com

MICHAEL MINGFU LING
Lorus Therapeutics, Inc.
Sunnybrook Health Sciences Centre
2075 Bayview Ave.
Toronto, Ontario M4N 3M5
Canada
Phone: (416)-231-4270
E-mail: mling@lorusthera.com

YING LUO
Rigel, Inc.
772 Lucerne Drive
Sunnyvale, CA 94086
USA
Phone: (408)-617-8005
Fax: (408)-317-8006
E-mail: yluo@rigelinc.com

BRIAN H. ROBINSON
Department of Genetics,
The Research Institute,
The Hospital for Sick Children,
Toronto, Canada

TATIANA I. SAMOYLOVA
Scott-Ritchey Research Center
College of Veterinary Medicine
Auburn University
Auburn, AL 36849
USA
Phone: (334)-844-5951
Fax: (334)-844-5850
E-mail: samoiti@auburn.edu

JÖRG A. SCHENK
Max Delbrück-Center
for Molecular Medicine
(MDC)
Robert-Rössle-Str. 10
13122 Berlin-Buch
Germany
Phone: +49-30-9406 2662
Fax: +49-30-9406 3895
E-mail: jschenk@mdc-berlin.de

PAUL D. SIEBERT
Gene Cloning and Analysis Group
Clontech Laboratories, Inc.
1020 East Meadow Circie
Palo Alto, CA 94303
USA
Phone: (650)-424-8222
Fax: (650)-354-0776
E-mall: paul_siebert@BD.com


PETER S. SILVERSTEIN
University of Kanses
Department of Pharmaceutical
Chemistry
2095 Constant Avenue
Lawrence, Kansas 66044
USA
Phone: (785)-864-4138
Fax: (785)-864-5736
E-mail: psilvers@mail.ukans.edu


KATHLEEN SIMPSON
Department of Neurology
Institute of Neurological Sciences
Southern General Hospital,
University of Glasgow
1345 Govan Road
Glascow, G51 4TF
Scotiand
Phone: 44 (0) 141 201 2465
Fax: 44 (0) 141 201 2515
E-mall: gora20@udcf.gla.ac.uk

BRUCE F. SMITH
Scott-Ritchey Research Center
College of Veterinary Medicine
Auburn University
Auburn, AL 36849
USA
Phone: (334)-844-5587
Fax: (334)-844-5850
E-mail: smithbf@auburn.edu


GIN WU
Cyberdent, Inc.
100 Galli Dr, Sulte 9
Novato CA 94949
USA
Phone: (415)-883-0484
Fax: (415)-883-3037
E-mail: cyberdent.com


YORK Y. ZHU
Gene Cloning and Analysis Group
Clontech Laboratories, Inc.
1020 East Meadow Circle
Palo Alto, CA 94303
USA
Phone: (650)-424-8222
Fax: (650)-354-0776
E-mall: yyzhu@clontech.com

# PCR and cDNA Cloning Techniques

# Strategies for cDNA Cloning and Mapping RNA Transcripts

PETER S. SILVERSTEIN, SHILPA J. BUCH, and R. CURTIS BIRD

## ▓ Introduction

The mapping of RNA transcripts is of fundamental importance to research in molecular biology. Thus, the determination of transcript boundaries and splicing patterns is most often the first task to be performed after the initial discovery of a transcription unit. The information obtained from transcript mapping studies can then be used to design other investigations concerning the relevant gene. Assuming that the transcript in question encodes a protein product, a definitive identification of structural motifs in the translation product can only be made after determination of the primary structure of the transcript. Attempts to define structural or functional motifs in proteins without mapping of the 5' and 3' ends of the transcript, as well as determining any splicing (including alternative splicing) patterns can lead to erroneous conclusions. Investigations such as

Peter S. Silverstein (e-mail: psilvers@mail.ukans.edu,
Tel: +1-785-8644138, Fax: +1-785-8645736)
University of Kansas, Department of Pharmaceutical Chemistry,
2095 Constant Avenue, Lawrence, Kansas 66044, USA
Shilpa J. Buch (e-mail: sbuch@kumc.edu)
Marion Merrel Dow Laboratory of Viral Pathogenesis, Department
of Microbiology, Molecular Genetics, and Immunology,
University of Kansas Medical Center, 3901 Rainbow Blvd., Kansas City,
Kansas 66160-7240, USA
R. Curtis Bird (✉) (e-mail: birdric@auburn.edu, Tel: +1-334-8445951,
Fax: +1-334-8445850)
Department of Pathobiology, Auburn University, Auburn,
Alabama 36849, USA

Springer Lab Manual
R.C. Bird, B.F. Smith (Eds.) Genetic
Library Construction and Screening
© Springer-Verlag Berlin Heidelberg 2002

the identification of promoters or enhancers also require that the primary structure of the transcript be previously determined.

There are laboratory guides to molecular biology that are excellent sources of information regarding some of the techniques used for mapping RNA transcripts (Ausubel et al. 1987; Sambrook et al. 1989). This chapter is not an attempt to replace those guides, rather, it is an effort to integrate some of this information and combine it with practical advice gleaned from the authors' personal experiences in this area. An overall strategy as to approaching the problem of transcript mapping is presented, and the technical details concerning each phase of the strategy are described. This presentation will certainly be useful to those with little experience in the evaluation and characterization of new RNA transcripts, but it should also prove to be of utility to investigators with more experience in the area.

## Outline

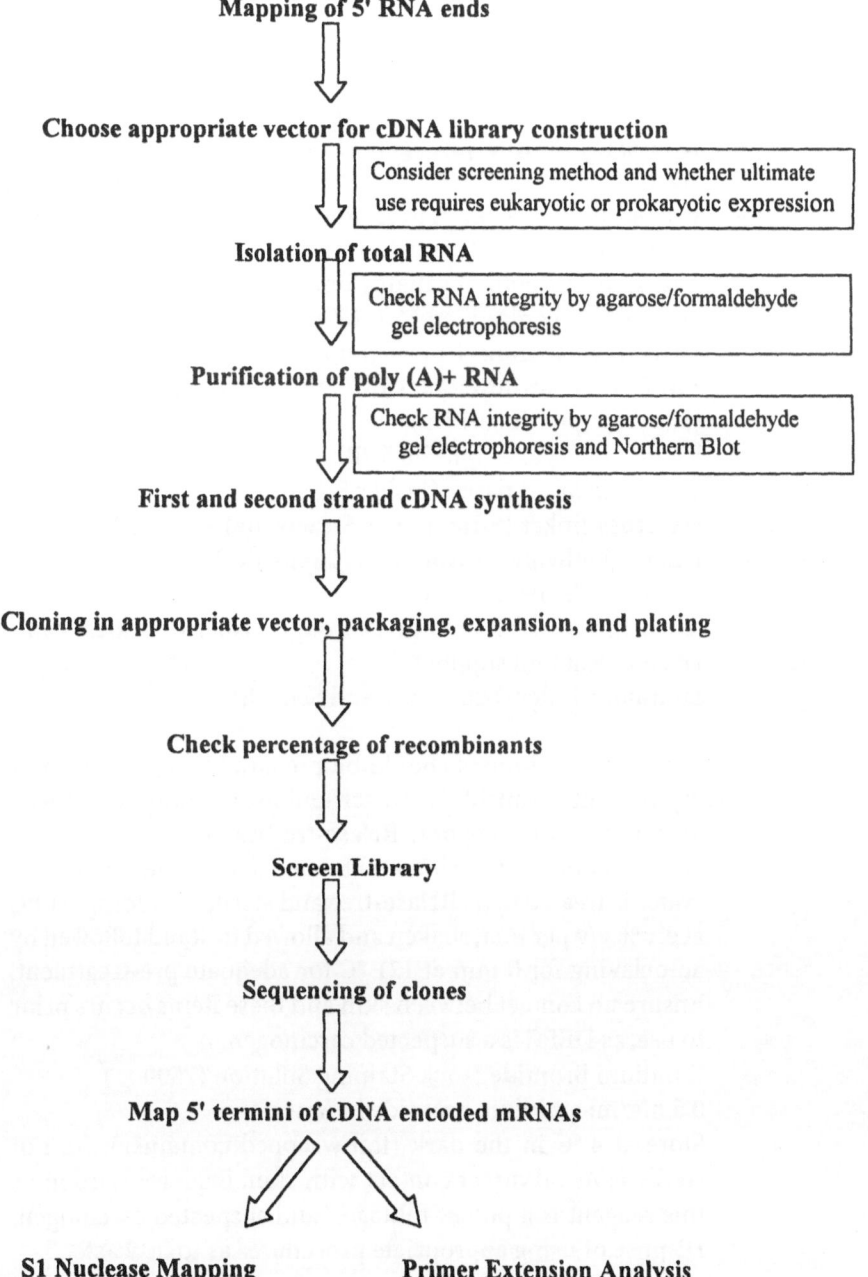

**Mapping of 5' RNA ends**

**Choose appropriate vector for cDNA library construction**

> Consider screening method and whether ultimate use requires eukaryotic or prokaryotic expression

**Isolation of total RNA**

> Check RNA integrity by agarose/formaldehyde gel electrophoresis

**Purification of poly (A)+ RNA**

> Check RNA integrity by agarose/formaldehyde gel electrophoresis and Northern Blot

**First and second strand cDNA synthesis**

**Cloning in appropriate vector, packaging, expansion, and plating**

**Check percentage of recombinants**

**Screen Library**

**Sequencing of clones**

**Map 5' termini of cDNA encoded mRNAs**

**S1 Nuclease Mapping**        **Primer Extension Analysis**

Fig. 1. Mapping of RNA transcripts and cDNA cloning strategy

## ▦ Materials

All standard laboratory supplies and reagents should be of reagent grade or higher quality of manufacture and can be obtained from either Fisher Scientific or Sigma Co.

- oligo(dT)-cellulose (Sigma Co.)
- DNase/RNase-free glycogen (Roche Molecular Biochemicals)
- Plastic centrifuge tubes in 15, and 50-ml sizes (Corning Plastics through Fisher Scientific)
- guanidinium isothiocyanate (Sigma Co.)
- enzymatically modified M-MLV (Moloney murine leukemia virus) reverse transcriptase (Superscript, Invitrogen)
- AMVRT (avian myeloblastosis virus) reverse transcriptase (Superscript, Gibco/BRL)
- RNase H (Superscript, Gibco/BRL)
- Whatman 3MM paper (Fisher Scientific)
- UV cross-linker (Stratalinker, Stratagene)
- DEPC (diethylpyrocarbonate) (Sigma Co.)
- Parafilm (Fisher Scientific)
- X-ray film (Kodak obtained through Fisher or a local camera/medical film supplier)
- Luminous Ruler (Schleicher and Schuell)

All buffers and solutions should be prepared using the highest quality reagents available, with scrupulous attention to RNase-free technique and using only RNase-free water.

- DEPC (diethylpyrocarbonate) treated and sterile water
  Water is treated to be RNase-free and sterile by adding DEPC at 0.1 % v/v per liter, shaken and allowed to stand followed by autoclaving for 0 min at 121 °C for adequate pre-treatment. Ensure no contact between skin and these items occurs prior to use, as DEPC is a suspected carcinogen.
- Ethidium Bromide Stock Staining Solution (2500 × )
  0.5 mg/ml ethidium bromide in water
  Store at 4 °C in the dark (foil-wrapped container). Do not sterilize. Avoid direct contact with skin. Exercise caution as this reagent is a potent mutagen and suspected carcinogen. Dispose of using appropriate procedures as toxic waste.

- 2 × SSC
  300 mM NaCl
  30 mM sodium citrate, pH 7.0
- Alkaline Lysis Solution
  0.5 M NaOH
  1.5 M NaCl
- Neutralization Solution
  1 M Tris-HCl, pH 8.0
- SM media
  0.1 M NaCl
  0.01 M $MgSO_4$-$7H_2O$
  2 % w/v gelatin
  20 mM Tris-HCl, pH 7.5
- Guanidinium Cell Lysis Solution
  4 M guanidinium isothiocyanate
  25 mM sodium citrate pH 7.0
  0.5 % Sarcosyl
  0.1 M β-mercaptoethanol

## Procedure

### Establishing and maintaining an RNase-free environment

For the methods discussed in this chapter, the establishment and maintenance of an RNase-free environment are absolutely essential. Standard recommendations include diethyl pyrocarbonate (DEPC) treatment of solutions, and baking glassware in a dry oven at 400 °F (204 °C) for a minimum of 2 h. Pipette tips that are certified to be RNase-free can be obtained from several suppliers. Stocks of pipette tips and microcentrifuge tubes, as well as stocks of DEPC-treated water and other solutions, should be set aside and used exclusively for RNA manipulations. Most microcentrifuge tubes do not need to be autoclaved with DEPC, as the heat generated during the manufacturing process is sufficient to inactivate nucleases. If the tubes are to be stored in beakers, use glass beakers with aluminum foil covers that have been baked. Hands, whether gloved or not, should not be inserted into containers containing microcentrifuge tubes. Rather, tubes should be poured out of the beaker or bag onto a clean surface and immediately capped. Like microcentrifuge tubes, disposable sterile 15-

and 50-ml centrifuge tubes can also be assumed to be RNase-free if unopened (Corning or Fisher Scientific). Fingers are a major source of RNase contamination. Gloves will minimize this source of contamination, but gloves need to be changed frequently. Using gloves that have been in extensive contact with surfaces such as doorknobs, door handles on freezers, is not much of an improvement over not using gloves. Attention should also be paid to anything that comes in contact with the Eppendorf tubes used for RNA manipulation. Finally, a good rule of thumb is that any plasticware that may have come in contact with anything that may contain RNase (e.g., a pipette tip that accidentally touched the outside of a bottle) should be discarded.

### RNA preparation

Before proceeding with library construction, it is important to determine that the tissue or cell line from which RNA will be isolated expresses the gene of interest. This can be confirmed by either Northern or Western blot analysis of the cell line or tissue from which RNA will be isolated for library construction (Ausubel et al. 1987; Sambrook et al. 1989).

Isolation of intact RNA from cultured cells is generally straightforward. The protocol of Chirgwin et al. (1979) has proven to be effective in our hands. Guanidine isothiocyanate is a strong denaturant of proteins, and thus is a very effective inhibitor of RNases. This is particularly important during initial lysis or if RNase-rich tissues are employed. The RNA obtained from this procedure is of very high quality, and can be used for any purpose. Several companies make solutions containing guanidium hydrochloride or guanidine isothiocyanate that are used in one step RNA isolation procedures (e.g., Tel-Test, Friendswood, TX or Invitrogen). These solutions work well for isolating RNA that will be used for Northern analysis. We have not tried to use RNA samples prepared in this manner for cDNA library construction, S1 nuclease analysis, or primer extension. The OD 260/280 ratios of RNA samples prepared with the one-step methods (OD 260/280 ratio <2.0) indicate that they are not as pure as RNA samples prepared by the method of Chirgwin et al. (1979). After isolation of total RNA, the integrity of the RNA should be checked by denaturing formaldehyde/agarose gel electrophoresis before an attempt is made to isolate poly(A)$^+$ RNA from the sample (Ausubel et al. 1987; Sambrook et al. 1989).

The integrity of total RNA samples can be evaluated by several methods. The easiest method is to analyze the RNA by electrophoresis on an agarose-formaldehyde gel just as for a Northern blot (Ausubel et al. 1987; Sambrook et al. 1989). Load 10–20 μg of total RNA for each sample, ensuring that an equal mass is loaded in each lane, and an additional lane containing an appropriate amount of an RNA molecular weight marker is included. After electrophoresis, the gel should be stained with an ethidium bromide solution (0.5 mg/ml). After destaining in several changes of four gel volumes of double distilled H$_2$O for a minimum of 2 h total, the 28S and 18S ribosomal RNA bands should be clearly visible. The ratio between the ethidium bromide stained 28S and 18S rRNA bands should be approximately 2:1 when compared under ultraviolet light illumination. This is because these two molecules should be produced stoichiometrically (in a 1:1 ratio), the amount of ethidium bromide bound is proportional to molecular length, and 28S rRNA is approximately twice the length of 18S rRNA. If this ratio is at or below approximately one (i.e., the bands are of approximately equal intensity), it can be safely assumed that significant RNA degradation has taken place and that the sample is not suitable for isolating poly(A)$^+$ RNA for use in first strand synthesis. Alternatively, after electrophoresis, one might proceed to transfer the RNA to a nylon membrane and hybridize the bound RNA with a probe for the gene of interest. The resulting bands should be sharp, and should not be smeared downward indicating that some degradation has occurred. Not only will this ascertain the quality of the RNA, but it will also provide some indication as to the abundance of the transcript(s) of interest.

After confirming that the total RNA is intact, poly(A)$^+$ RNA should be isolated. The method used for many years was to use column affinity-chromatography to bind the poly(A)$^+$ fraction of a total RNA preparation (Ausubel et al. 1987; Sambrook et al. 1989). Total RNA is loaded onto a column of oligo(d)T-cellulose and the poly(A)$^+$ fraction bound to the matrix. After several washes, the poly(A)$^+$ fraction is eluted in a low-salt buffer. Although this method yields good results, recent years have seen the advent of simpler, more user-friendly methods. These newer methods eliminate most of the problems associated with column chromatography (i.e., slow flow rates, clogging, matrix preparation, etc.) though. These systems are sometimes a bit more expensive than column chromatography, however, they are well worth the investment for cDNA library construction. Column chromatography only becomes cost effective if a large number of poly(A)$^+$ RNA samples have to be prepared. Whatever system is used, the integrity of the poly(A)$^+$ RNA must be confirmed. This can be performed by Northern blot analysis of a 1 µg sample of the poly(A)$^+$ RNA using a β-actin or GAPDH probe.

## cDNA library construction

cDNA library construction represents a major investment in time and effort, however, a high quality library can be an invaluable long-term resource for an investigator. Because there are several good kits available for cDNA library construction, it is suggested that this approach be followed. Even with a kit, however, library construction is neither simple nor trivial. This is balanced by the availability of technical support from the manufacturer and kits purchased from reputable firms generally yield excellent results when high quality RNA is used. Although kits tend to be more expensive initially, from a practical point of view, it is only worth investing the time to optimize conditions for library construction reagents in instances where a large number of libraries need to be constructed.

Most of the cDNA library construction kits commercially available utilize modifications of the method of Gubler and Hoffman (1983). Although there are a variety of kits that are commercially available, the basic protocols are very similar. Briefly, a primer or linker/primer is annealed to poly(A)$^+$ RNA and first-strand synthesis is performed with either avian myeloblastosis virus reverse transcriptase (AMVRT) or Moloney murine leukemia virus reverse transcriptase (MMLVRT). Second-strand synthesis is then performed using a mixture of RNaseH and DNA polymerase I. The RNaseH nicks the RNA in the first strand cDNA/RNA hybrid that is the product of first-strand synthesis. The RNA fragments are then able to serve as primers for DNA polymerase, which synthesizes the second strand. The resulting cDNA is then treated with T4 DNA polymerase to produce blunt ends on which to ligate either linkers or adaptors. Adaptors are designed to have "sticky" (single-stranded) ends specific for a particular restriction enzyme site, whereas linkers require digestion with a restriction enzyme, along with some method to protect the cDNA from digestion (e.g., methylation). If linker or linker primers are used, the double-stranded cDNA will be cleaved with the appropriate restriction enzyme. Because of the higher efficiency of cloning with adaptors, most companies have switched to their use, wherever possible. The resulting double-stranded cDNA molecules are size-fractionated and ligated into a lambda-based cloning vector. The ligation mixture is then packaged, after which the library is plated onto an appropriate *E. coli* host strain.

Although the availability of kits for the construction of cDNA libraries has simplified the process considerably, a number of factors should be considered when choosing such a kit. The method used to screen the library will determine the nature of the lambda cloning vector. If screening with an antibody is the method of choice, then a lambda expression vector must be chosen. However, if a nucleic acid probe is to be used for screening, then other considerations will ascend in priority. For example, directional cloning is possible with some vectors, allowing determination of which end of the cDNA represents the 5' end of the mRNA. This becomes important when trying to obtain nearly full-length cDNAs.

The method by which inserts can be obtained from purified plaques using different phage vectors is another factor that should receive consideration. Traditionally, after obtaining a pure plaque, inserts from lambda clones have been isolated directly from large scale restriction digests of phage preparations. This can be problematic because obtaining high-quality preparations of lambda DNA can be difficult. The problem is compounded when attempting to isolate smaller inserts, as these represent a smaller percentage of the recombinant phage genome. In recent years, as a solution to this problem, several lambda-based in vivo excision systems have been developed. These systems utilize a helper phage that is co-infected into an appropriate strain of *E. coli* along with a plaque-purified lambda clone. Through a recombination event, a phagemid is excised from the lambda clone and is recovered in plasmid form, thus obviating the need for DNA preparations from phage. If this route is decided upon, the features and properties of the excised phagemid should be assessed to see if that particular vector will be useful for later applications. For example, some phagemids contain eukaryotic promoters and this is a necessary feature for studies involving transfection of eukaryotic cells. In general, consider the applications downstream of cDNA library construction for which clones will be used, and then choose the system that best meets those needs.

## Results

### Tips on cDNA library construction and screening

Although most instruction manuals that are included with cDNA library construction kits are sufficient for their purpose, a few procedures that are not described in kit manuals may be of help to investigators constructing their first cDNA libraries. The first, and most fundamental advice, is to read the manual thoroughly and plan ahead. Reactions are performed at different temperatures, and incubators and water baths should be prepared ahead of time. For each different reaction temperature, there should be one incubator or water bath set, stabilized, and at that temperature before starting the procedure. When setting up reactions, be aware of the next step in the protocol and make sure the necessary equipment and reagents are available. Except when noted below, perform all reactions under the conditions specified by the manual as the kits have already been optimized for the reactions as described.

One modification that we would suggest concerns the ligation of the double-stranded cDNAs to the lambda vector. The manufacturers' protocols suggest that a specific ratio of cDNA to vector be used in the ligation mix. Our suggestion is to set up three ligations, one of which follows the ratio(s) suggested by the manufacturer. The other two ligations should utilize the same amount of vector with adjustments to the amount of double-stranded cDNA added to the reaction. Using three times as much double-stranded cDNA as recommended for one reaction, and one third as much for a third reaction, should provide a productive range of vector/insert ratios. These ligations should then be packaged individually, and the resulting libraries evaluated separately for the number of recombinants and range of insert sizes in the library. In all other respects, it is suggested that investigators adhere closely to the recommendations of the manufacturer.

The packaging reaction providing the best results should then be amplified. Amplification should be performed only once to avoid biases in relative cDNA abundance. One large amplification should be all that is needed for most purposes. Store some aliquots at 4 °C for working stocks, and also store aliquots at –80 °C. The stocks in the freezer will not only provide long term storage, but will also serve as a back-up in the event of some mishap with the stocks at 4 °C.

Library screening is the next task to be performed after amplification. Useful protocols for screening cDNA libraries can be found in several molecular biology laboratory manuals (Ausubel et al. 1987; Sambrook et al. 1989). These manuals describe methods for titering and plating phage libraries. Schleicher and Schuell (1995) also produces an excellent pamphlet that includes information on library screening, along with information on Northern, Southern, and Western blotting. The following should help those with little or no experience in this endeavor. Various membranes are available for library screening; we recommend the use of either nylon or reinforced nitrocellulose membranes. Although the nylon membrane is a bit tougher, the reinforced nitrocellulose membranes usually produce lower levels of background, and are strong enough to maintain their integrity if treated with a minimum of care. Plain (i.e., non-reinforced) nitrocellulose membranes are not recommended because of their fragility. Filters should never be handled with bare hands, and even handling with gloved hands should be avoided. Membrane forceps (i.e., those with blunt tips) are available from many scientific supply houses, and these should be the only tool used in manipulating membranes.

Once the plaques have grown up, it is best to chill the agar plate for 2–4 h at 4 °C before proceeding with plaque lifts; this makes the top agarose less fragile. Before laying a membrane on a plate, label the edge of the membrane with the plate number using an alcohol-proof marker. The membrane also needs to be marked so it is known which side has plaques. To lay the membrane on the plate, hold the circular membrane with forceps at opposite ends so that the membrane is held horizontally. The membrane is then formed into a "U" shape, and the bottom of the "U" is placed on an imaginary line representing the diameter of the plate. Releasing the two ends of the membrane from the forceps will allow the membrane to unfold relaxing evenly onto the top agar in the plate. Make sure there are no air bubbles; if the filter does not lie on the agar properly then remove it and use another filter. The filter should wet in a few seconds and should be left in contact with the top agarose for approximately 2 min for the first lift; subsequent lifts can be left on the agarose longer (up to 10 min). Before removal from the plate, the membrane should be marked asymmetrically with a syringe filled with India ink and fitted with a 20–24 gauge needle. This will allow the positive signals on the autoradiogram made from the filter to be related to the position of plaques on the plate. Before starting plaque lifts, make sure that the flow of ink from the syringe is at an appropriate level by making mock stabs into an agar plate. There should be only enough ink to leave a small mark in the agar. Large amounts of ink will diffuse and alignment of the autoradiogram with the plate will be difficult. It is important to remove the filter in one smooth motion to avoid smearing plaques. After plaque lifts have been made, the plates should be sealed with Parafilm and stored at 4 °C.

After removal from the plate, each filter should be treated with alkaline lysis solution to denature the DNA. The filter should then be neutralized by sequential treatments with excess neutralization solution and finally 2 × SSC, as described (Schleicher and Schuell 1995). This is easily accomplished by making small stacks (i.e., 3–4 pieces) of Whatman 3MM paper soaked in the appropriate solution and placing the stacks on a sheet of cellophane wrap on a laboratory bench. Each filter is then individually transferred from stack to stack as it goes stepwise through the denaturation/neutralization process. The side of the filter that lies on the agarose (i.e., the side containing the plaques) should always be facing up to avoid smearing and cross contaminating subsequent filters. The use of the 3MM paper assures that the plaques will not be smeared by excess liquid, yet it provides sufficient solution for effective processing. If many filters are to be processed, the solution in each stack can be replenished by adding solution to the top of the stack and allowing it to be absorbed. Care should be taken to avoid pools of liquid on top of each stack, as this can result in smearing.

After denaturation and subsequent neutralization, the DNA should be fixed to the membrane. This can be accomplished by baking at 80 °C or by UV cross-linking. UV cross-linking also enhances the signal, so our filters are first cross-linked and then baked at 80 °C. Although the use of both measures may not be absolutely necessary, the extra time and effort required is trivial.

For hybridization of the filters, the easiest method is to perform prehybridization and hybridization reactions in a beaker that is of slightly greater diameter than the filter. An aliquot of prehybridization solution is added to a beaker and filters are added to the beaker one at a time, making sure that each filter is properly wet prior to the addition of the next filter. The beaker is then covered with cellophane wrap and incubated at the appropriate temperature. At the end of the prehybridization incubation, fresh hybridization fluid including the radioactively labeled probe, is added to a second beaker. The filters in the prehybridization fluid are then individually transferred to this beaker. The beaker is then covered with cellophane wrap and incubated in a water bath at the temperature appropriate for the probe in use (45–55 °C for low to standard stringency and 65 °C for high stringency washes). Prehybridization and hybridization reactions should be incubated at the same temperature.

After hybridization and washing, the filters are air dried and placed in cellophane wrap. At this point, for ease of identification, the holes in the filter used for orientation can be marked with a permanent marker. A length of cellophane wrap is unrolled and laid upon a laboratory bench and filters are placed on the wrap and then covered with another layer of wrap. After exposure to X-ray film and developing of the autoradiogram, the film must then be oriented to both the filters and the plates from which the lifts were made. The film is first oriented with regard to the filter. This can be accomplished by marking the cellophane-wrapped filters in one of several ways. There are a few biotechnology companies that market luminous rulers, or other luminous markers that can be placed in the cellophane portion of the filter/cellophane assembly. Alternatively, many discount stores carry stickers for children that are luminous, will expose the film and can be used to orient the film to the filter/cellophane assembly. If using these stickers, first do a test exposure with a small piece of film. Whatever method is used, try to use at least three small orientation markers for each filter/cellophane package. After orientation of the film to the filters, the positions of the holes in the filters should be marked on the film. The plates are then oriented to the film and positive plaques are selected and marked on the back of the plates.

Methods for picking positive plaques from a plate are varied, and a matter of personal choice. Our preferred method is to use a glass Pasteur pipette from which the tapered end has been removed by scratching with a diamond pencil and breaking it. A small pipette bulb is then placed on this end, air is expelled from the bulb and the other (wrong) end of the pipette is used to cut a large plug containing the plaque of interest from an agar plate. After insertion of the pipette end into the agar, suction from the pipette bulb is used to remove the plug, which is then expelled into 1 ml of SM media to elute the selected phage stock for the next round of screening.

After several rounds of screening and purification, at progressively lower numbers of phage per plate, purified isolates will be obtained. Plates containing putative isolates can be assessed by plaque lifts and hybridization to the probe used in the initial screen. If the plaques on the plate are indeed purified, then there should be no plaques that do not hybridize to the probe. If there are plaques on the plate that do not hybridize to the probe, at least two more rounds of plaque purification are needed. One of the tricks used to obtain pure isolates is to plate plaques at a lower density in later rounds of screening. This facilitates the isolation of individual plaques from a plate. If the library has been constructed in a vector compatible with an in vivo excision system as described above, a plasmid containing the cDNA can be readily obtained from plaque-purified phage by co-infection of *E. coli* with the plaque-purified phage and a helper phage. If a vector was used in library construction that is not compatible with an in vivo excision system, then a stock of phage DNA must be prepared and the cDNA insert excised and manually cloned into a plasmid vector. Obviously, the in vivo excision systems are easier to use and more convenient. However, when screening pre-made libraries, there may not be a choice as to the vector used for library construction.

## Characterization of cDNA clones

After obtaining cDNA inserts from a library, they should be completely sequenced and the expression of the encoded mRNAs characterized. Sequencing will yield data on splicing patterns, the size of the open reading frame, and other features that affect expression. Expression can be assayed by labeling the cDNA insert and using it as a probe in Northern blots (Ausubel et al. 1987; Sambrook et al. 1989). If the expression of the mRNA is inducible or is tissue-specific, then this attribute can also be evaluated by the use of appropriate mRNA samples. It is also important to characterize the size of the mRNA(s) identified by the cDNA insert(s) to ensure that the cDNA insert encodes the mRNA observed in the first Northern and the results are not due to shared homology between the products of different open reading frames. If Western analysis was used to identify the cell line or tissue expressing the gene of interest, then it is important to confirm that the antibody used to initially detect the encoded protein, is also capable of detecting the protein encoded in the cDNA. Some phage vectors used in cDNA library construction will express fusion proteins under the control of an inducible promoter (generally the *lac* promoter). For proper expression, it is critical that the insert be cloned in-frame with the prokaryotic portion of the fusion. Alternatively, after sequencing, the insert can be transferred to a plasmid expression vector and the identity of the encoded protein confirmed by Western analysis.

## S1 Nuclease and primer extension analyses

The information obtained from analysis of cDNA is often sufficient for the needs of most investigations. Analysis of cDNAs yields a wealth of information regarding both the mRNA and the putative protein product. However, the first few nucleotides of the mRNA will often not be represented in the cDNA, thus the start site of transcription may remain unknown. This information can be determined by a combination of S1 nuclease and primer extension analyses. Both of these techniques are in wide use, and complete descriptions and protocols are available in popular laboratory manuals, as well as in the primary literature (Ausubel et al. 1987; Sambrook et al. 1989; van Santen 1991, 1993; Silverstein et al. 1995, 1998). S1 nuclease analysis is used to determine the start site of transcription, or the presence of a 3' splice acceptor. Primer extension analysis is used to confirm the results obtained from S1 nuclease analysis, and to confirm that the putative start site determined by S1 nuclease analysis is the actual start site and not a 3' splice acceptor. Some of the important factors to consider in these reactions are parameters such as temperature of hybridization, temperature of the primer extension reaction, and the amount of S1 nuclease used in digestion. Any, or all of these parameters can be modified, and the appropriate modifications will differ when using different primer/RNA or probe/RNA combinations. A good starting point for both S1 nuclease analysis and primer extension analyses are the conditions cited in either Sambrook et al. (1989) or Ausubel et al. (1987).

Adherence to a few general guidelines will considerably ease the work involved in transcript mapping. For all of these procedures, obtaining high quality RNA is critical. Since these procedures involve multiple steps and are performed over extended periods of time, proper planning in the execution of these procedures is of utmost importance. Because of the expense in terms of time and material, any intermediate steps should be checked for proper results. If unfamiliar with mechanical procedures (e.g., plaque lifts, pulling agarose plugs, etc.), set up a mock experiment which can be used to practice the mechanical skills necessary. It is important to remember that few, if any, of these procedures are "cookbook" molecular biology. For S1 nuclease and primer extension analyses, optimization of conditions for each primer/RNA or probe/RNA combination is important for obtaining best results. As in most areas of science, proper planning will save time, money, and materials.

## References

Ausubel FM, Brent R, Kingston RE, Moore DD, Seidman JG, Smith JA, Struhl K (1987) Current protocols in molecular biology Wiley, New York

Chirgwin JJ, Przbyla AE, MacDonald RJ, Rutter WJ (1979) Isolation of biologically active ribonucleic acid from sources enriched in ribonuclease. Biochemistry 18:5294

Gubler U, Hoffman BJ (1983) A simple and very efficient method for generating cDNA libraries. Gene 25:263–269

Sambrook J, Fritsch EF, Maniatis T (1989) Molecular cloning: a laboratory manual. Cold Spring Harbor Laboratory Press, Cold Spring Harbor

Schleicher, Schuell (1995) Blotting, hybridization, and detection: an S&S laboratory manual. Schleicher and Schuell, Keene, NH

Silverstein PS, Bird RC, Van Santen VL, Nusbaum KE (1995) Immediate-early transcription from the channel catfish virus genome: characterization of two immediate-early transcripts. J Virol 69:3161–3166

Silverstein P, Van Santen VL, Nusbaum KE, Bird C (1998) Expression kinetics and mapping of the thymidine kinase transcript and an immediate-early transcript from channel catfish virus. J Virol 72:3900–3906

van Santen VL (1991) Characterization of the bovine herpesvirus 4 major immediate-early transcript. J Virol 65:5211–5224

van Santen VL (1993) Characterization of a bovine herpesvirus 4 immediate-early RNA encoding a homolog of the Epstein-Barr virus R transactivator. J Virol 67:773–784

## Suppliers

Clontech Laboratories, Inc., 1020 East Meadow Circle, Palo Alto, CA 94303–9605, USA (Tel.: +1-800-6622566, Fax: +1-650-4248222, www.clontech.com)

Corning Inc., P.O. Box 5000. Corning, New York 14830, USA (Tel.: +1-607-9744640)

Fisher Scientific, 2000 Park Lane Drive, Pittsburgh, Pennsylvania 15275-9943, USA (Tel: +1-800-7667000, www.fishersci.com)

Invitrogen Corporation, 1600 Farady Avenue, Po.B. Box 6482 Carlsbad, California 92008, USA (Tel.: +1-800-8286686, Fax: +1-800-3312286, www.lifetech.com)

Schleicher and Schuell, Inc, P.O. Box 2012, Keene, New Hampshire 03431, USA (Tel.: +1-800-2454024, Fax: +1-603-3523810)

Sigma Co., P.O. Box 14508, St. Louis, Missouri 63178, USA (Tel.: +1-800-3255832, www.sigma-aldrich.com)

Stratagene, 11011 North Torrey Pines Road, La Jolla, California 92037, USA (Tel.: +1-800-4245444, Fax: +1-619-5355400, www.stratagene.com)

Tel-Test, Inc., 1511 County Road 129, P.O. Box 1421, Friendswood, Texas 77546, USA (Tel.: +1-281-4822672, Fax: +1-800-6310600)

# Cloning PCR Products

WEIWEN JIANG and BRUCE F. SMITH

## ▓ Introduction

Cloning PCR products into a stable vector is often desirable for
subsequent analysis such as high quality DNA sequencing, hybridization studies, expression, modification or further subcloning.
A wide variety of strategies have been employed and are available
in commercially supplied kits (see Page 30 for a listing of manufacturers of PCR cloning kits or vectors). These approaches
include simple but relatively inefficient blunt ligation schemes
that rely on the 5' to 3' proofreading activities of some enzymes
such as *Pfu* polymerase (Costa and Weiner 1994). These enzymes
can be used to create blunt-ended PCR products or to remove the
3' overhang generated by *Taq* polymerase, creating "polished
ends". A modification of this approach uses the restriction
endonuclease *Srf 1* to cleave the vector DNA and then supplies *Srf
1* in the ligation buffer to provide a higher steady-state concentration of digested vector, which forces the reaction to proceed in
the direction of ligation of the insert, as the ligated product is not
digestible by *Srf 1* (Simcox 1992). Other recent innovations
include the addition of 5' AATTC tails to PCR primers followed
by amplification in a mix containing phosphorothioate dGTP.
The PCR product is then digested with Exonuclease III, which
removes bases from the 3' end of the strand until it encounters
the protected G, leaving an AATT "sticky end", perfect for ligation
into a vector's *Eco R1* site. Yet another method uses the Lambda
phage recombination/integration system rather than ligation as

W. Jiang, B.F. Smith
Scott-Ritchey Research Center, College of Veterinary Medicine,
Auburn University, Auburn, Alabama 36849, USA

Springer Lab Manual
R.C. Bird, B.F. Smith (Eds.) Genetic
Library Construction and Screening
© Springer-Verlag Berlin Heidelberg 2002

a mechanism for cloning (reviewed in Landy 1989). In this approach, attB sites are incorporated into the PCR primers and attP sites are added to the vector. When the PCR product, vector, and Lambda phage integrase (*Int*) are combined, the PCR product is inserted into the vector in a site-specific manner. The same system can then be used with Lambda phage excisionase (*Xis*) to remove the insert for transfer to a different vector. Finally, restriction endonucleases, such as Eam 1104 I, which cuts at a defined distance from its recognition site (but not within its site) are used in PCR primers to allow specific "sticky ends" to be produced (Padgett and Sorge 1996). However, most of these approaches are either inefficient, or require the use of proprietary vectors, enzymes, or reagents and are often costly.

One approach lends itself well to duplication in the laboratory using commonly available reagents, and hence is extremely cost effective. This approach relies on the fact that *Taq* polymerase has a nontemplate-dependent activity which normally adds a single nucleotide (almost always A) to the 3' ends of all amplicons (Clark 1988). The efficiency of direct cloning of PCR products can be improved by this one base overhang, which facilitates ligation when the complementary T base is added to the cloning vector. This cloning strategy is called T/A PCR cloning. If *Vent* and *Pfu* polymerases are employed in the PCR reaction, there are no 3'A overhangs because of the 3'→5' exonuclease activity of these polymerases. The blunt-ended PCR products derived from these PCR reactions can be cloned into blunt-ended vectors. Alternatively, a 3' A-overhang can be added by incubation with *Taq* polymerase at the end of the PCR cycles.

There are many commercially available TA cloning kits, however, it is relatively simple and inexpensive to make your own T/A cloning vector. Generating a T/A cloning vector in your own laboratory may also give you the advantage of accelerating experimental progress. For example, if further study in a mammalian expression system is needed and you have an eukaryotic expression vector, you can find or make a blunt-end restriction site in the multiple cloning site. Using the following protocols, any vector can be made into a T/A cloning vector. After cloning your gene of interest, you can directly express your gene without having to clone into a T/A cloning vector and subsequently subclone it into an expression vector. Because T/A cloning is bidirectional, restriction digestion patterns or sequencing must be used to check for insert orientation. Further in vitro and in vivo expression can then be carried out. The following is a basic protocol to generate a home-made T/A cloning vector.

## ▨ Outline

The entire protocol of generating T/A cloning is outlined in Fig. 1.

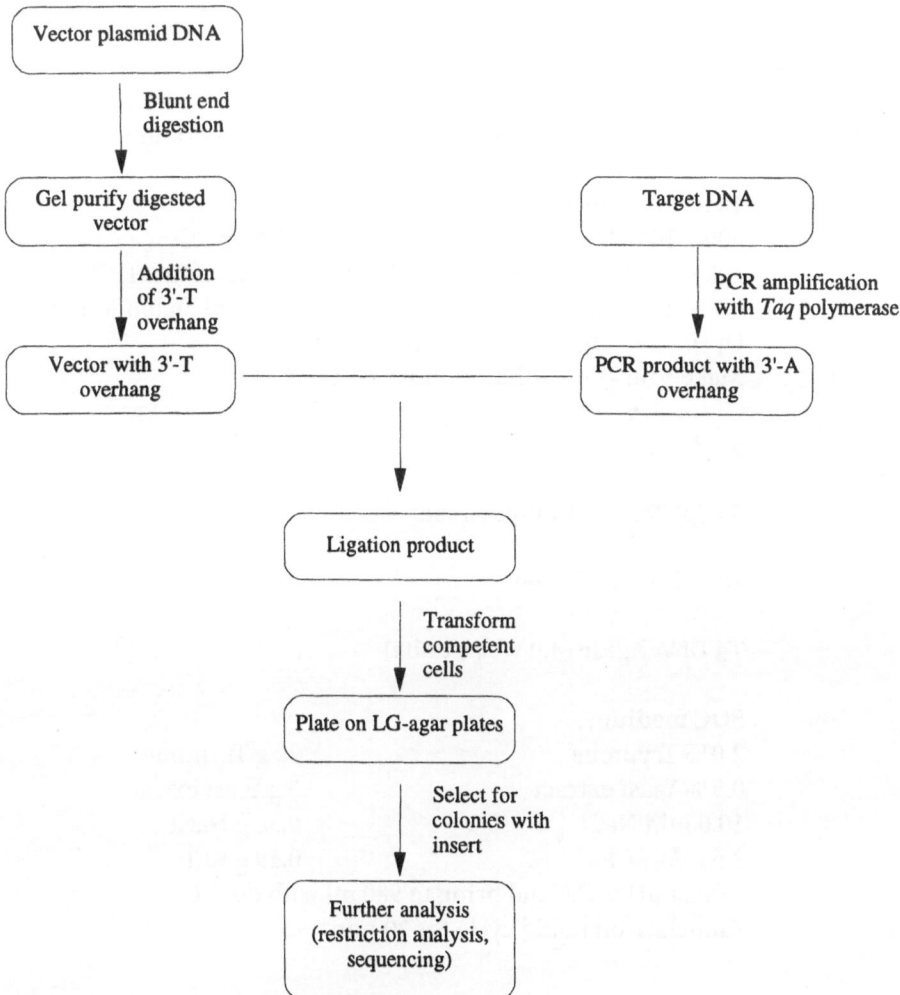

**Fig. 1.** Outline of T/A PCR cloning procedures

## ▦ Materials

Equipment — 75 °C heat block
— PCR thermocycler
— Agarose gel electrophoresis box
— Water bath (14 and 42 °C)
— 37 °C incubator
— UV spectrophotometer
— 37 °C rotary shaker
— 5 µg vector DNA (e.g., pBlueScript, pUC19, M13 mp18, etc.)

Reagents **TE buffer, pH 8.0**
10 mM Tris HCl, pH 8.0        10 ml of 1.0 M Tris HCl
1 mM EDTA, pH 8.0             2 ml of 0.5 M EDTA
Make 1 liter stock solution, store at room temperature

**10 × PCR buffer**
500 mM KCl                              37.27 g KCl
100 mM Tris HCl, pH 9.0 (at 25 °C)    100 ml of Tris HCl
0.1 % Triton X-100                     0.001 ml Triton X-100
Optimized $MgCl_2$ concentration
Make 1 ml stock solution, store at –20 °C

**5 mM dTTP**

**5 U/µl *Taq* DNA polymerase**

**10 × ligation buffer**

**T4 DNA ligase** (4.0 Weiss units)

**SOC medium**
2.0 % Tryptone              20 g Tryptone
0.5 % Yeast extract         5 g Yeast extract
10.0 mM NaCl               0.58 g NaCl
2.5 mM KCl                 0.19 g KCl
Adjust pH to 7.0 and bring to 980 ml with $ddH_2O$
Autoclave on liquid cycle for 20 min

10.0 mM $MgCl_2$-$6H_2O$ (autoclaved)
20.0 mM glucose (filter-sterilized)
Total volume of 1 liter, store at room temperature or 4 °C

**Competent cells**

**LB-agar medium**

| | |
|---|---|
| 1.0 % Tryptone | 10 g Tryptone |
| 0.5 % Yeast extract | 5 g Yeast extract |
| 1.0 % NaCl 10 g NaCl | |

Make to 950 ml, adjust pH to 7.0 with NaOH and bring to 1 liter
Add 15 g agar and autoclave at liquid cycle for 20 min

**Antibiotics**

Additional reagents for restriction endonuclease digestion, agarose gel electrophoresis, PCR, DNA fragment ligation and DNA transformation.

## Procedure

### Prepare blunt-ended vector DNA

1. Digest plasmid DNA vector with a restriction endonuclease that yields blunt ends. Examples of common enzymes are *EcoRV, Pvu II, Sca I*, etc.

2. Examine the plasmid digest by gel electrophoresis to check for correct digestion

3. Purify the DNA by electrophoresis and subsequent gel extraction, phenol/chloroform extraction and precipitate with absolute ethanol or a DNA purification column. Centrifuge at 4 °C at maximum speed on a bench top centrifuge, wash with 70 % ethanol, air dry 10 min at room temperature and resuspend in a suitable volume of TE buffer, pH 8.0.

### Addition of T-overhang

1. Set up the following reaction to generate T-overhangs:
   5 μg blunt-ended vector DNA
   10 μl 10 × PCR buffer
   20 μl 5 mM dTTP
   1 μl *Taq* DNA polymerase (5 U/μl)
   Add sterile ddH$_2$O to 100 μl

2. Incubate in 75 °C heat block for 2 h.

3. Use immediately, if possible, otherwise store at −85 °C.

### PCR amplification of target DNA

1. PCR-amplify the desired target DNA under optimized conditions. At the end of the reaction add an extra 5–15 min extension cycle at 70–75 °C to be sure all PCR fragments have A-overhangs. Additional dATP may be added prior to this step if PCR reaction is not in the logarithmic phase.

2. Recover the PCR fragment by gel electrophoresis, phenol/chloroform extraction or a PCR purification column and precipitate with ethanol, centrifuge at top speed (4 °C) and dissolve in 25 μl TE, pH 8.0. Determine PCR fragment concentration by optical density at 260 nm using a UV spectrophotometer.

**Note.** If *pfu* or *Vent* polymerases are used in the PCR reaction, *Taq* polymerase can be added at the end of the PCR reaction and extended at 70 °C

### Ligation of vector and PCR amplicon

1. Set up the following ligation reaction:

   | | |
   |---|---|
   | Fresh PCR product | X μl |
   | 10 × ligation buffer | 1 μl |
   | T-overhanged vector | 2 μl |
   | 10 mM ATP | 0.5 μl |
   | Sterile H$_2$O | up to a total of 9 μl |
   | T4 DNA ligase | 1 μl |
   | Total volume | 10 μl |

Insert and vector molar ratio can range from 1:1 to 1:3. The following equation can be used to determine how much PCR product is needed to ligate into the vector with a 1:1 molar ratio:

$$X \text{ ng PCR product} = \frac{(\text{bp of PCR product}) \, (\text{ng of the vector})}{(\text{bp the vector})}$$

If 1:N molar ratio is needed, simply multiply X by N to get the amount needed for ligation.

2. Incubate the above ligation reaction at 14 °C for at least 4–6 h (preferably overnight).

**Note.** If you are unable to proceed to the next step immediately, store the reaction at –20 or –85 °C until ready for transformation.

### Transformation of ligation reaction

Use competent cells for your transformation. **Note.** If blue-white colony selection is employed, you need to check if the cell contains the *lac* repressor. When a cell contains the *lac* repressor, IPTG inducing is required, while without repressor, the product may be expressed in the absence of IPTG.

1. Thaw competent cells on ice (keep on ice all the time).

2. Carefully pipette 50 µl of competent cells into pre-chilled vials. **Do not pipette up and down.**

3. Pipette 2 µl ligation reaction directly into competent cells and mix by stirring gently with the pipette tip. Incubate on ice for 30 min. Left over competent cells need to be quick frozen in dry ice/ethanol and stored at –80 °C.

4. Heat shock for exactly 45 s in the 42 °C water bath. Do not shake or mix. Remove the vials immediately from water bath and incubate on ice for 2 min.

5. Add 150 or 250 µl of SOC medium (room temperature) to each tube.

6. Shake the vials at 37 °C for 1 h at 225 rpm in a rotary shaker

7. Spread 100 µl from each transformation vial on separate, labeled LB-agar plates containing appropriate antibiotics and selectable markers.

   **Note.** If *lacZ* is the marker, spread the transformed cells on X-gal-containing agar plates. Be sure to include IPTG if you are using *lac* repressor-containing competent cells.

8. Place all the plates in a 37 °C incubator overnight to let colonies grow. After incubation at 37 °C overnight, incubate at 4 °C for 2–3 h to allow for proper color development if blue-white colony selection is used.

9. Select for positive clones

   Pick colonies that may contain inserts for further plasmid isolation and restriction analysis.

## Results

The benefit of T/A cloning is the generation of stable plasmid clones for PCR products. The selection of positive clones is critical, because of the possibility of PCR error, especially when generating long PCR products. This problem can be reduced if *Vent* or *Pfu* polymerase are used. However it must be noted that these polymerases do not add a terminal A, so *Taq* polymerase must be used to do so. After plasmid isolation and restriction endonuclease digestion, sequencing of the insert is highly recommended before further studies. Once a positive clone is identified, further studies such as hybridization, subcloning to expression vectors, etc., can be carried out.

## Troubleshooting

If you do not obtain the results you expected, use the following chart to troubleshoot your experiment.

| Problems | Answers |
|---|---|
| No colonies obtained from transformation | Try to use a plasmid DNA as a positive control for your transformation reaction. Your cells may not be competent.<br>Check the concentration of antibiotic on your plates.<br>Always spread your transformation on pre-warmed plates. |
| Colonies do not have insert | The single 3' T-overhangs are not added correctly or have been degraded. Use freshly made vectors and avoid repeated freeze/thaw cycles. Also check for vector self-ligation by including a ligation without insert. |
| Lots of colonies on plates | There might be some contamination of your reaction. Try to have a negative control transformation with no plasmid as a control for contamination. |
| Very few colonies | Majority of colonies are blue or light blue with very few white colonies. Make sure you are using Taq polymerase in your PCR reaction.<br>Try to use fresh PCR products, because the single 3' A-overhang may be degraded over time. Also try to avoid a gel purification step of your PCR product, since it may remove the A-overhang.<br>If your insert is less than 500 bp, you may have light blue colonies. Check some of the light blue colonies to see if they contain insert.<br>The molar ratio of your PCR reaction and vector is incorrect. Check the molar ratio using the equation provided above.<br>Too high a salt content in the ligation reaction may inhibit ligation. Use no more than 2–3 µl of the PCR mixture in the ligation reaction. |
| Only white colonies | Check to see if you have X-Gal. If you are using competent cells with lac Z repressor, be sure you have IPTG on your plate. |

## ▮ Application

### Application of this technique to blunt cloning any DNA fragment

The strategy of T/A PCR cloning can also be utilized to increase the efficiency of blunt-ended ligation. Blunt-end digested DNA can be added with a 3'-A and cloned into a vector with a 3'-T overhang. After the selection of the vector with the insert, sequencing or restriction mapping must be performed to check for the direction of the insert because the ligation is bidirectional.

## ▮ References

Clark JM (1988) Novel non templated nucleotide addition reactions catalyzed by procaryotic and eukaryotic DNA polymerases. Nucleic Acids Res 16:9677–9686

Costa GL, Weiner MP (1994) Polishing with T4 or Pfu polymerase increases the efficiency of cloning of PCR fragments (journal article). Nucleic Acids Res 22(12):2423

Landy A (1989) Dynamic, structural, and regulatory aspects of lambda site-specific recombination. Annu Rev Biochem 58:913–949

Padgett KA, Sorge JA (1996) Creating seamless junctions independent of restriction sites in PCR cloning. Gene 168(1):31–35

Simcox TG, Marsh SJ, Gross EA, Lernhardt W, Davis S, Simcox ME (1992) SrfI, a new type-II restriction endonuclease that recognizes the octanucleotide sequence, GCCCGGGC. Gene 109(1):121–123

## ▮ Suppliers

### Manufacturers of commercial PCR cloning products

Amersham Pharmacia Biotech Inc., 800 Centennial Ave., Piscataway, New Jersey 08855–1327, USA (Tel.: +1-732-4578000, Fax: +1-800-FAX-3593, Toll Free: 800-526-3593, www.apbiotech.com)

CLONTECH Laboratories Inc., 1020 E. Meadow Circle, Palo Alto, California 94303–4230, USA (e-mail: orders@clontech.com, Tel.: +1-650-4248222, Fax: +1-800-4241350, Toll Free: 800-662-2566, www.clontech.com)

Genpak Inc., 25 E. Loop Rd., Stony Brook, New York 11790–3350,
USA (e-mail: k.cox@genpakdna.com, Tel.:: +1-631-4446625,
Fax: +1-631-4446626, Toll Free: 800-385-9248,
www.genpakdna.com)

Invitrogen Corp., 1600 Faraday Ave., Carlsbad, California 92008,
USA (e-mail: tech × service@invitrogen.com,
Tel.: +1-760-6037200, Fax: +1-760-6037201,
Toll Free: 800-955-6288, www.invitrogen.com)

Life Technologies Inc., 9800 Medical Center Dr., Rockville,
Maryland 20849–648, USA (GIBCO BRL) (e-mail:
info@lifetech.com,Tel.: +1-301-6108000, Fax: +1-800-3312286,
Toll Free: 800-828-6686, www.lifetech.com/)

Novagen Inc., 601 Science Dr., Madison, Wisconsin 53711, USA
(e-mail: novatech@novagen.com, Tel.: +1-608-2386110,
Fax: +1-608-2381388, Toll Free: 800-526-7319,
www.novagen.com)

Promega Corp., 2800 Woods Hollow Rd., Madison, Wisconsin
53711, USA (e-mail: custserv@promega.com,
Tel.: +1-608-2744330, Fax: +1-608-2772601,
Toll Free: 800-356-9526, www.promega.com)

Roche Molecular Biochemicals, Div. of Roche Diagnostics,
9115 Hague Rd., P.O. Box 50414, Indianapolis, Indiana 46250,
USA (Tel.: +1-800-4285433, Fax: +1-800-4282883,
Toll Free: 800–262–1640, biochem.roche.com)

Sigma-Aldrich, 3050 Spruce St., St. Louis, Missouri 63103, USA
(Tel.: +1-314-7715750, Fax: +1-800-3255052,
Toll Free: 800-325-3010, www.sigma-aldrich.com)

Stratagene, 11011 N. Torrey Pines Rd., La Jolla, California
92037–1073, USA (e-mail: tech × services@stratagene.com,
Tel.: +1-858-5355400, Fax: +1-858-5350045,
Toll Free: 800-424-5444, www.stratagene.com)

Fermentas Inc., 7520 Connelley Dr., Ste. A, Hanover, Maryland
21076, USA (e-mail: info@fermentas.com, Tel.: +1-905-6900800,

Fax: +1-800-4728344, Toll Free: 800-340-9026,
www.fermentas.com)

Gemini Biotech, Sid Martin Biotechnology Institute, 12085
Research Dr., Alachua, Florida 32615, USA
(e-mail: gembio@geminibio.com, Tel.: +1-904-4622249,
Fax: +1-904-4622283, Toll Free: 800-561-3454,
www.geminibio.com)

Genomics One Corp., 3030 Le Carrefour Blvd., Ste. 902, Laval,
QC H7T-2P5 Canada (e-mail: info@genomicsone.com,
Tel.: +1-450-6884499, Fax: +1-450-6889100,
Toll Free: 877-684-3637, www.genomicsone.com )

# Site Directed Deletion, Insertion, and Substitution using PCR

MICHAEL MINGFU LING and BRIAN H. ROBINSON

## ▓ Introduction

During the past decade, the approaches to DNA mutagenesis have undergone fundamental changes. Early techniques used to obtain a mutant involved screening numerous living organisms or cells for naturally existing mutants, often following exposure to a mutagen. Since DNA is reproduced in vivo quite faithfully according to its template and most errors are efficiently detected and reversed by DNA repair systems, naturally occurring mutations are rare and thus are difficult to isolate. Subsequently, with advancing technologies, DNA mutants could be created in vitro at will (Smith 1985). However, the methods were inconvenient, because the thermolabile enzymes and single-stranded DNA templates were used often with only one round of DNA synthesis. Recently, significant improvements have been made in using thermolabile polymerases as enzymes and double-stranded DNA as templates, by integrating strong selection methods, e.g., Unique Site Elimination based on restriction digestion at a site unique to the wild type (Deng and Nickoloff 1992), into the mutagenesis methods.

    The development of PCR-based methods has revolutionized the technologies by which mutants are obtained, allowing the use

M.M. Ling (✉) (e-mail: mling@lorusthera.com,
Tel.: +1-416-2314270
Lorus Therapeutics, Inc., Sunnybrook Health Sciences Centre,
2075 Bayview Ave., Toronto, Canada M4N 3M5
B.H. Robinson
Department of Genetics, The Research Institute,
The Hospital for Sick Children, Toronto, Canada

Springer Lab Manual
R.C. Bird, B.F. Smith (Eds.) Genetic
Library Construction and Screening
© Springer-Verlag Berlin Heidelberg 2002

of thermostable enzymes together with double-stranded DNA templates and improving the efficiency of mutagenesis. Currently, a variety of PCR-based methods are available to meet different experimental needs. In general, all of these methods are convenient, reliable and effective for DNA mutagenesis.

All PCR-based methods can be classified into random mutagenesis and site-directed mutagenesis (SDM). Random mutagenesis implies that the position or the type of mutations is created randomly over a region ($\geq 4$ amino acid residues) and is thus not predefined. On the other hand, site-directed mutagenesis implies that the position or the type of mutations is predefined and is created precisely at a specific site or sites ($\leq 3$ amino acid residues) with a specific mutation. In SDM PCR methods, a mutation is introduced by a mutagenic primer. Because of its simplicity, the primer has become a convenient source for introducing mutations. Depending on the design of a mutagenic primer, the outcome of such SDM mutations can be a deletion, an insertion or a point mutation (substitution). In this chapter, the use of site-directed mutagenesis techniques directed towards generating all of these mutation categories will be the focus.

## Outline

One of the popular PCR methods for SDM is the Megaprimer PCR, since it is the simplest and most versatile method. Three primers (two outside primers and one middle mutagenic primer) are required for this method. Usually, a small number of mismatches, in a mutagenic primer, are tolerable for such a primer binding to its template, so that mutations are introduced precisely by the mutagenic primer into a newly synthesized PCR product. By designing a mutagenic primer in a desired position, mutations can be introduced anywhere in the final PCR product. The mutagenic primer can also be designed to encode any, or a combination, of the following mutations: a deletion, an insertion or a point mutation, all of which are usually fewer than 10 bp. In the Megaprimer method, a double-stranded plasmid containing a gene (wild-type) can be conveniently used as a template without further manipulation.

In the initial version of this Megaprimer method, two separate PCRs are required. The first PCR is designed to produce a short PCR product using the mutagenic primer, the matching outside primer and the wild-type template. This short PCR product is termed megaprimer, for the reason that one strand of the megaprimer is then fully extended, using wild-type templates, to form a single-stranded full-length mutant template. In the second PCR, the mutant template is amplified by two outside primers to generate final mutant products. Here, we present an improved version of the Megaprimer PCR method (Ling and Robinson 1995) that offers not only the ultimate convenience but also the high efficiency in comparison to the initial version of the Megaprimer method. More importantly, this improved method allows the amplification of a final mutant product at least 2 kb in size using Pfu DNA polymerase, or 5.3 kb using Taq DNA polymerase, via a megaprimer as large as 1.3 kb. Thus this improved method is particularly useful to mutagenize a large gene-cassette without any pre-existing unique restriction sites in the region of interest, which is often the case.

This improved method is dubbed as One-Step Three-stage Efficient PCR, and is abbreviated as "One-STEP" (Fig. 1). In this method, an operator can set up a three-stage PCR (three consecutive and automatic cycling) in one tube (all three primers and all components for the PCR are added in the same tube at once, prior to initiation of the reaction). By programming the software, the reaction is carried out automatically without any human intervention. Since the PCR is programmed into three stages with individually optimized parameters for each stage, different products are formed in different stages, i.e., megaprimer in stage one, mutant template in stage two, final mutant product in stage three.

The function of stage one is to form megaprimer carrying mutations, using wild-type templates, mutagenic primers and matching outside primers. In comparison to a standard PCR, a shorter time of denaturation and annealing is used in this method to decrease the heating time, thus to minimize the damage to the template and to the DNA polymerase, and to facilitate the eventual formation of final mutant products in the following stages. The shorter extension time (5 s+60 s/kb of the megaprimer), used in this stage, is designed to be adequate for the amplification of megaprimers, and to be inadequate for the amplification of undesirable full-length "short-circuited" products derived from two outside primers and wild-type templates. This shorter extension

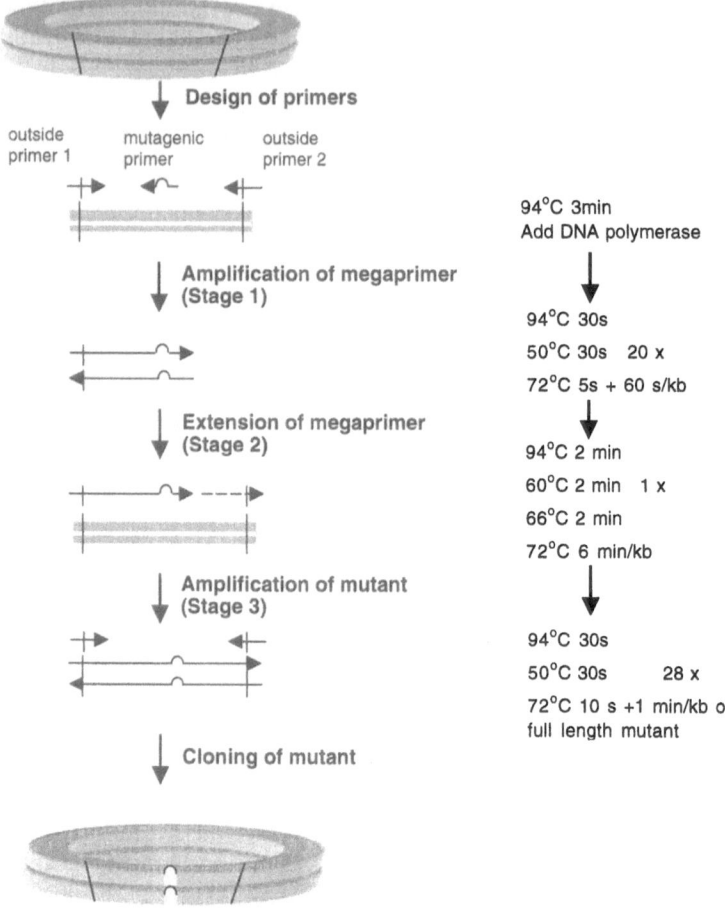

**Fig. 1.** The improved Megaprimer method. *Grey shaded circles* and *bars* represent wild-type plasmid DNA as template. *Vertical lines* indicate the positions of unique restriction digestion sites. *Half circles* represent the mutation sites. *Short horizontal arrows* represent outside primers. *Long horizontal lines with arrows* represent strands of DNA in the 5' terminus to 3' terminus direction

time is crucial in encouraging formation of the megaprimer and discouraging formation of the full-length product (wild type). The function of stage two is to produce a full-length mutant template containing mutations by extending one of the two strands of the megaprimer formed in stage one, along wild-type templates. Usually, the melting temperatures for two outside primers are around 55–58 °C and that for megaprimers, around 70 °C. Thus, the higher annealing temperatures (60 and 66 °C) are used in this stage to prevent the two outside primers from binding to the template, and at the same time to allow megaprimer binding to the template, thus minimizing the background of wild types and maximizing the formation of mutant templates. The annealing is also divided into two temperatures to avoid the abrupt ramping in the temperature, thus to maintain the annealed megaprimer-template complex in the temperature transition period and to facilitate megaprimer extension. Since ample time is used for each denaturation (2 min), annealing (60 °C 2 min, 66 °C 2 min) and extension (72 °C 6 min/kb), the optimal cycle number for this stage is one, which was experimentally determined. The function of stage three is to amplify the mutant template obtained in stage two, using two outside primers. As in stage one, a shorter time (30 s) of denaturation and annealing is employed for stage three. About 60 s/kb (length of the full length mutant) of extension time is used for adequate amplification of mutant templates to obtain final mutant products.

Since three primers are present in the same tube, a potential problem is that the mutagenic primer may only react with one of the outside primers to form predominantly the short PCR product, i.e., the megaprimer. To avoid this problem, the mutagenic primer should be added in much lower amounts than that of the outside primers. As a result, the mutagenic primer is completely incorporated into the product by the end of stage one.

## ▨ Materials

The following products and their suppliers were used in our experiments. It is understood that these products can be substituted with similar ones from other suppliers, without affecting the success of the procedure.

- Pfu DNA polymerase (Stratagene) for preparing mutants up to 2 kb; combination of Taq DNA polymerase (Life-Technologies, Qiagen, or Stratagene) and Taq Extender additive (Stratagene) for preparing larger mutants (up to 5.3 kb). Pfu PCR buffer (10 ×) as supplied by Stratagene: 200 mM Tris-HCl (pH 8.8), 100 mM KCl, 100 mM $(NH_4)_2SO_4$, 20 mM $MgSO_4$, 1% Triton X-100, 1 mg/ml nuclease-free bovine serum albumin. Taq PCR buffer (10 ×) as supplied by Life

Technologies: 200 mM Tris-HCl (pH 8.4), 500 mM KCl. with MgCl$_2$ 50 mM separately provided.
- dNTPs, 100 mM each (Amersham-Pharmacia or Life Technologies)
- Oligonucleotide primers are commercially available from many suppliers. They are resuspended to a concentration of 10 µM. The melting temperature (Tm) should be around 55–60 °C for all three primers.
- Thermocycler (e.g., Perkin-Elmer 9600) and PCR reaction tubes (thin-wall type preferred)
- Agarose gel solutions, 10 mg/ml ethidium bromide solution (Sigma) and electrophoresis apparatus
- System of DNA purification from gel, e.g., QiaXII kit from Qiagen or GeneClean kit from Bio101.
- DNA plasmid templates (wild type), prepared according to the standard miniprep procedure (Sambrook et al. 1989)
- Relevant restriction endonucleases (5–20 units/µl from New England Biolabs), T4 DNA ligase (e.g., from Life Technologies), or cloning kit (e.g., TA cloning kit from Invitrogen)
- *Escherichia coli* competent cells (Invitrogen)

## Procedure

### PCR

1. Set up PCR reactions on ice according to the following recipe:

| Component/reaction | Wild-type control | Megaprimer control | Mutant construction |
|---|---|---|---|
| | | Volume (µl) | |
| Pfu buffer (10 ×) | 5 | 5 | 5 |
| dNTP (10 mM) | 2 | 2 | 2 |
| Outside primer 1 (10 µM) | 2 | 2 | 2 |
| Outside primer 2 (10 µM) | 2 | 0[a] | 2 |
| Mutagenic primer (1.6 µM) | 0 | 2 | 2 |
| Plasmid (1 ng/µl) | 1 | 1 | 1 |
| Sterile distilled water | 38 | 38 | 36 |
| Total volume | 50 | 50 | 50 |

[a] **Note:** assuming megaprimer is amplifiable between outside primer 1 and the mutagenic primer.

2. For those thermocyclers that do not have a heated lid, add 50 μl light mineral oil (Fisher Scientific) to avoid condensation as a result of temperature cycling on the lid of a PCR tube. Add the thermostable polymerase after "hot-start", i.e., pre-heat the reaction mixture for 2 min at 94 °C, and then add 0.5 μl Pfu (5 U/μl) or Taq 5 U/μl and Taq Extender additive 5 U/μl.

3. Program the PCR machine using the following cycling parameters (three consecutively executed stages):

| Stage 1 (20 × ) → | Stage 2 (1 × ) → | Stage 3 (28 × ) |
|---|---|---|
| 94 °C 30 s | 94 °C 2 min | 94 °C 30 s |
| 50 °C 30 s | 60 °C 2 min | 50 °C 30 s |
| 72 °C 5 s+60 s/kb of | 66 °C 2 min | 72 °C 5 s+60 s/kb of |
| megaprimer | 72 °C 6 min/kb | full-length mutant |

As in any PCR, the reactions should be incubated at 72 °C for 10 min after thermal cycling. Finally, the completed reactions can be conveniently kept at 4 °C on the machine until the next step. Both the final 72 °C incubation and the 4 °C soak can be pre-programmed.

**Gel electrophoresis**

1. After the PCR, run the reactions on a 0.8 % agarose gel with 1 × TBE and 20 μg ethidium bromide in the gel.

2. After the electrophoresis, the gel is placed under UV illumination to visualize bands. Ensure that the correct sized mutant band is present in the lane loaded with "mutant construction" reaction.

3. Excise the mutant band from the gel with a blade. Around 0.5 μg of the mutant product should be present (see next step for quantitation). If necessary, repeat and replicate the above PCR to accumulate more products. Purify the DNA, using a kit such as QiaXII or GeneClean.

4. Quantify the purified DNA on an agarose gel using DNA quantitative standards in which bands have predetermined

quantities, such as the φx 174 RF DNA/HaeIII digest (a marker supplied by Life Technologies).

### Restriction digestion, ligation and transformation

1. To clone the mutant PCR product, a vector fragment and a mutant fragment, which is derived from the mutant PCR product, need to be prepared. To prepare the vector fragment, take 500–1000 ng of plasmid, which was used in the PCR step earlier, incubate it with two restriction enzymes, the sites of which are unique in the plasmid and are built in the mutant PCR product through primers. To prepare the mutant fragment, set up a digestion with 300–500 ng of the purified mutant PCR product using the same pair of restriction enzymes.

|  | Vector fragment reaction | Mutant fragment reaction |
|---|---|---|
| DNA | Wild-type plasmid (500–1000 ng) | Mutant PCR product (300–500 ng) |
| Enzyme buffer (10 ×) | 1 µl | 1 µl |
| Restriction enzyme 1 | 0.5 µl | 0.5 µl |
| restriction enzyme 2 | 0.5 µl | 0.5 µl |
| Sterile distilled water | To a total reaction volume of 10 µl | To a total reaction volume of 10 µl |

Incubate the reaction for 2–5 h at 37 °C.

2. Run the reaction on a 0.8 % agarose gel. Ensure that the digestion is complete by checking the vector reaction to visualize two distinct bands. One is the vector and the other is the wild-type insert which will be replaced by a mutant fragment. Excise the vector fragment and the mutant fragment. Gel purify as in point 3; gel quantify the bands as in point 4 (under "Gel electrophoresis"). Dilute the purified vector fragment and mutant fragment each into 10 ng/µl

3. Set up the ligation as follows.

|                          | Vector control | Mutant construction |
|--------------------------|----------------|---------------------|
| Ligation buffer (5 ×)    | 2 µl           | 2 µl                |
| Vector double digest fragment (10 ng/µl) | 4 µl | 4 µl      |
| Mutant double digest fragment (20 ng/µl) | 0 | x µl (mutant : vector molar ratio 3:1) |
| Ligase (10 U/µl)         | 1 µl           | 1 µl                |
| Sterile distilled water  | 3 µl           | To a total volume of 10 µl |

Incubate the ligation reaction at 22–25 °CC for 1 h (in order to carry on with the next step on the same day) or 16 °C overnight (in order to carry on with the next step the next day). For better ligation efficiency, incubate the ligation reaction on a thermocycler with a cycling profile of 16 °C for 1 min and then 30 °C for 1 min, for 4 h to overnight.

4. Use 1 µl of the completed ligation to transform bacterial competent cells, either commercially obtained or prepared in the laboratory (Hanahan 1983). Perform a control transformation with 1 ng of pUC 18 to confirm the expected transformation efficiency ($10^6$–$10^8$ transformants/µg DNA). The number of transformants from the "mutant construction" should be significantly higher than that from the "vector only" control, which should yield no or very few colonies.

5. The resulting transformants are individually characterized. Cultures are grown from the transformants and then plasmid DNAs are prepared. The miniprepped plasmid DNA is analyzed by restriction digestion using the same pair of restriction enzymes. Finally, plasmid DNAs with mutant fragments are sequenced to confirm the presence of the desired mutation and the absence of any unwanted mutations in the rest of the gene sequence. Readers are encouraged to refer to other chapters of the manual for tips on ligation and transformation.

## ▮ Troubleshooting

### No product is observed at all at the end of a PCR

If the wild-type control reaction also did not produce any bands on a gel, the problem reflects the deficiency or the difficulty with the PCR amplification in general, and it is not unique to the Megaprimer protocol. Potential sources for this problem include incomplete or defective PCR components, or incorrect concentrations or amounts of PCR components in the reaction, such as enzymes, primers, dNTPs or templates.

The cycling conditions may have been inappropriate. The cycling parameters, mentioned in the procedure section for the improved Megaprimer method, are provided as guidelines. However, the shorter denaturation, annealing and extension time for the first stage, the higher annealing temperature for the second stage and longer extension time for the third stage should be followed strictly. For different thermocyclers and different reaction volumes, the cycling parameters may differ from what we have suggested here. It is recommended that before using a new thermocycler and using a different reaction volume, the time and temperature parameters be first tested before performing the mutagenesis PCR.

Sometimes, the sequence of a template may possess a high GC content or contain a complex secondary structure that resists denaturation even under the PCR condition, which affects the formation of final PCR products. Relevant to this problem, alkaline and heat denaturation of the template and primers before a PCR, the use of denaturants such as DMSO, and/or high denaturation temperature can significantly improve the amplification of these difficult DNAs.

For certain large inserts, long PCR technology should be applied. For example, the Pfu or Vent can be used as a secondary enzyme in addition to Taq as the main enzyme, to increase the size of a PCR product. We have found that Taq Extender from Stratagene generally works well for both long and short PCR products. A smaller reaction volume and different buffer compositions (e.g., high pH and addition of denaturants) are often useful in obtaining long PCR products.

Finally, the design of primers is a key to the success of DNA mutagenesis (see Comments 2 and 3).

**The size of the PCR product is shorter than that of the expected mutant construct. In some cases, there is predominantly the shorter product with little expected full-length mutant construct**

If the short product is the same size as the megaprimer product in the "megaprimer control" PCR reaction, it is usually caused by the higher-than-suggested amount ratio of mutagenic primers vs. two outside primers. The amount of mutagenic primers should be significantly lower than that of the outside primers, as suggested in the procedure section.

The template-independent addition of a nucleotide (most often adenosine, or A) to the 3'-end of a product by Taq is often seen in a PCR product. If this terminal addition occurs to a great extent, in the megaprimer, it stalls the extension of the megaprimer. Thus, in the final PCR product, the short product is predominant. An interesting solution to this problem is to design the mutagenic primer in such a way that its 5'-end is immediately preceded by a T, thus any A added to the sequence of a megaprimer will not interfere with the subsequent extension of a megaprimer.

Megaprimers, being long, double-stranded DNAs, are often difficult to denature, anneal and extend to form the full-length mutant template. It is noted that the megaprimer can be extended more efficiently by lengthening the time of denaturation, annealing and extension, as is designed in, and recommended for, stage two of this improved megaprimer method.

**The size of the PCR products is correct; however, most clones obtained contain wild-type sequence, and not the desired mutation**

In this case, the conditions of the PCR reaction may have favored the formation of wild-type DNAs which have a similar or identical size to the mutant PCR product. One of the likely reasons is that too little mutagenic primer is used in the PCR. Alternatively, the template amount may have been too high. If more than 1 ng template is used, it may be necessary to use a higher amount of mutagenic primers (see also Comment 3, when higher amounts of templates are required, a remedy based on using ddNTP-treated restriction digestion fragments should be considered).

Another reason is that the extension time in stage one may be too long so that the wild-type full length product as well as the megaprimer is amplified. This extension time should be short enough to only effectively amplify the megaprimer.

### There is a spontaneous mutation in addition to the desired mutation in the obtained clone

Random mutations do occur in PCRs. One simple solution to this problem is to sequence three to five clones simultaneously. The average frequency of random mutations, in a PCR, is in the range of $10^{-4}$–$10^{-6}$ for Taq and $10^{-4}$–$10^{-6}$ for Pfu and Vent (Cline et al. 1996), which means that, for a gene insert of 500 bp in size, after finding a random mutation in one clone, it is very likely that the next clone would be free of random mutations.

If this problem is epidemic among several clones, measures can be taken to reduce spontaneous mutations. Taq DNA polymerase should be substituted with Pfu, since Pfu DNA polymerase produces 6–15-fold better fidelity than does Taq DNA polymerase (Cline et al. 1996) under the same reaction conditions. Fewer cycles in the third stage should also be considered to further reduce the error frequency, since amplification through many cycles accumulates more errors. Although the decrease in cycles may result in a decrease in the final PCR product, the use of a higher amount (>1 ng) of template will make it possible to obtain a large amount of product with a smaller number of cycles (see also Troubleshooting guide 3 regarding the amount of mutagenic primer). When higher amounts of templates are employed, a dideoxynucleotide triphosphates (ddNTP)-blocked restriction digestion fragment, derived from the wild-type plasmid, can be used as a template in addition to the wild-type templates (see Comment 3).

Together with other colleagues, we have also found that the tendency of spontaneous undesired mutations is also gene- and sequence-dependent. Some DNA constructs may encounter more sequence fidelity problems than do other constructs.

**Although the correct sized product is obtained from the PCR, few or no transformants contain the correct mutant insert**

Cloning PCR products are often tricky. Although the PCR product is observable on a gel, often the synthesis is incomplete at both ends of a product, yielding staggered ends, i.e., sticky ends. These incomplete ends, when near the restriction site, often make the digestion impossible. Consequently, the ligation of a vector fragment with a mutant fragment is affected, causing the transformation to fail. Similarly, the imperfect ends also prevent the TA cloning and the blunt-end cloning of the products (see earlier chapters for cloning of PCR products). The remedy to imperfect ends is to prolong the final 72 °CC extension step of a PCR from a conventional 5–10 min to 30–60 min, thus forcing complete synthesis at both ends of a PCR product. Sometimes, when restriction sites are built too close to the end of a PCR product, those sites will digest with difficulty even when two ends of a PCR product are complete. Therefore, extra bases on the 5' side of the restriction site should be added (see appendices of the New England Biolabs catalogue).

Another likely reason is that the purified PCR product, even gel-purified, may contain residual Taq DNA polymerase and trace amounts of dNTPs. These contaminants are therefore present in the reaction of restriction enzyme digestion of a PCR product. While the PCR product is being digested, the resulting restriction ends are immediately modified by the residual Taq enzyme and dNTPs. The modification of restriction ends thus leads to a low efficiency or even failure of a ligation. The solution to this problem is the thorough cleaning of a PCR product before setting up the restriction digestion. It was demonstrated that three rounds of phenol/chloroform were necessary for removing Taq DNA polymerase (Bennett and Molenaar 1994). We have also found that DNAs purified by GeneClean kits resulted in successful ligation and transformation in refractory cases when Qiagen kits were used. An alternative to a thorough cleaning of PCR products is the use of some Taq DNA polymerase inhibitors, which inactivate the enzyme. As a matter of fact, these inhibitors are so effective that they can be added to a completed PCR reaction, without further purification. Such inhibitors are available commercially, for example, Taq-Quench from Life Technologies.

## ■ Comments

### 1. Other major SDM PCR methods

In addition to the Megaprimer PCR method, several other major types of SDM PCR methods are available, which will be briefly described here. The Connection PCR (Fig. 2A) is the ligation of two PCR products, one or both carrying mutations. To incorpo-

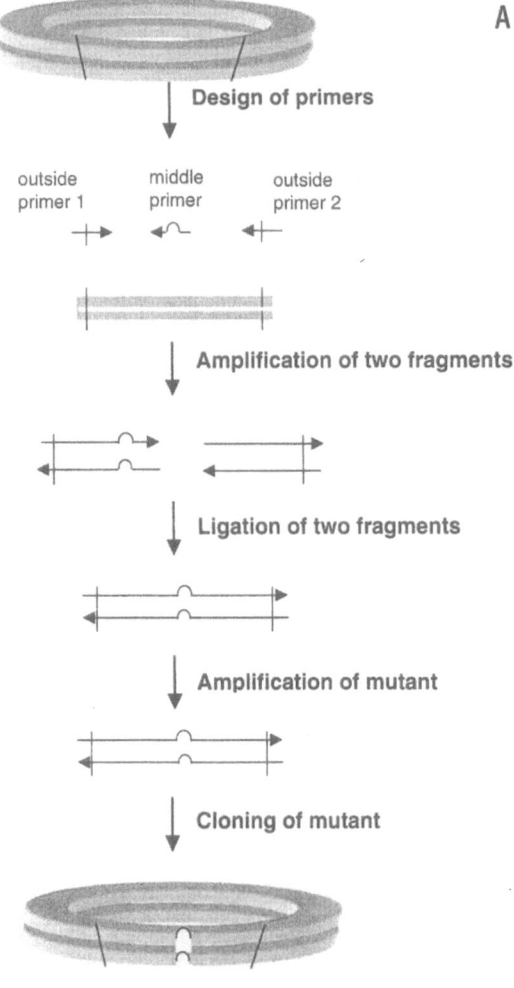

**Fig. 2.** Several major PCR mutagenesis methods. **A** Connection PCR; **B** overlap extension PCR; **C** inverse PCR

rate the mutation, the gene insert is first "split up" into two fragments at the desired mutation site, by performing two separate
PCRs, each using a pair of primers. Thus, four primers (two outside primers and two middle primers) are used in these two
PCRs. One or both middle primers can contain the pre-designed
mutation. To complete the procedure, the two fragments are subsequently re-connected at the desired mutation site by either in
vitro ligation or in vivo recombination (Ling and Robinson
1997). The second PCR method is the Overlap Extension method
(Fig. 2B), in which two pairs of primers are used, similar to the

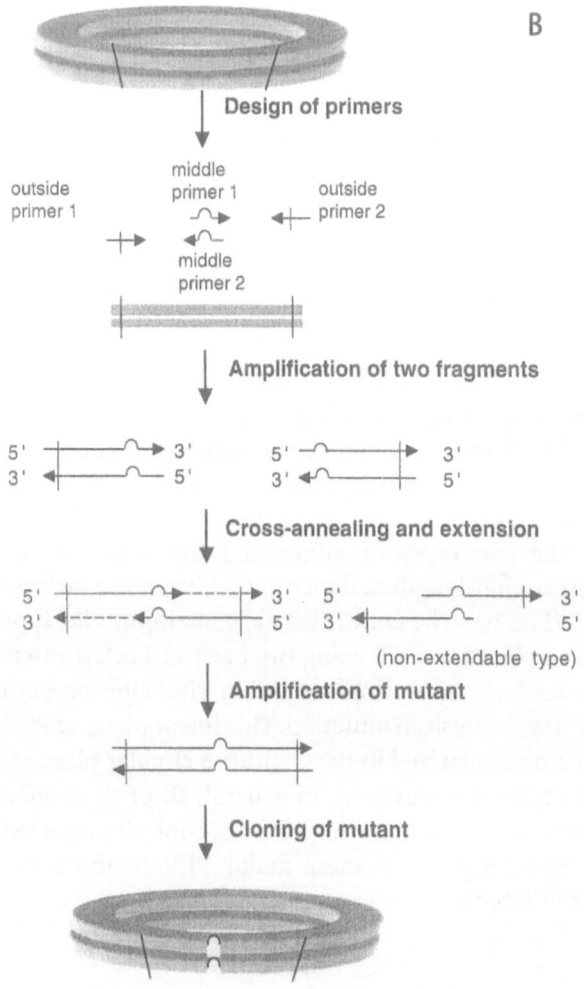

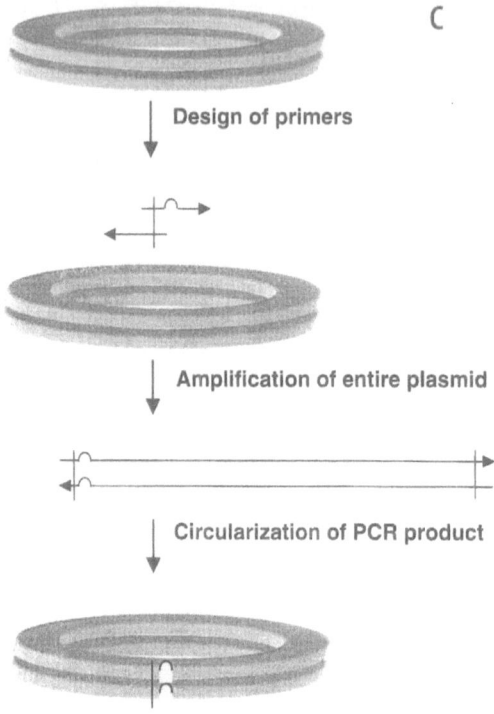

Connection PCR. The middle mutagenic primers must be designed to be completely, or for the most part, overlapped with each other and both should contain the designed mutation. Two separate PCRs are carried out in parallel to produce two fragments. Then these two fragments are joined by cross-annealing. Only one of the two types of annealed DNAs can then be extended to form a full-length mutant product. The third method is Inverse PCR (Fig. 2C). The entire plasmid containing wild-type gene insert is amplified by PCR using two back-to-back primers which are located at the desired mutation site. One or both primers can carry the desired mutation. This linear plasmid PCR product is then recircularized to reconstitute a circular plasmid, which now contains the mutation. In general, all of these PCR methods mentioned are reliable and effective for site-directed mutagenesis. Some aspects of these major PCR methods are compared in the Table 1.

**Table 1.** Comparison of several PCR mutagenesis methods

|  | Improved Megaprimer | Connection or Overlap Extension | Inverse |
|---|---|---|---|
| No. of PCR required | 1 | 2 | 1 |
| End product | A larger mutant insert | A mutant insert | A linear mutant plasmid |
| Mismatch position in the primer | Middle | Middle or 5'end for connection PCR; middle only for Overlap Extension | Middle or 5'end |
| Need ligation with insert? | Yes, its cloning is the same as that for the wild-type insert | Yes, its cloning is the same as that for the wild-type insert | No, but still need ligation or in vivo Recombination to circularize the product |
| Advantage | Convenient, fast, effective, and versatile | Good efficiency, relatively fast, effective | Mutant insert already in a linear plasmid, fast |
| Disadvantage | Needs careful setup of parameters | Takes longer procedural time and efforts | Long PCR techniques are likely needed |

## 2. Primer design

The success of SDM largely relies on the optimal design of a primer. Primers in general should have perfect homology to the binding site of the template. Nonetheless, mutations (deletion, insertion and point mutation) can be "smuggled" into the middle of a sequence of a mutagenic primer. For Megaprimer and Overlap Extension PCRs, mismatches should be placed in the middle of a primer, rather than 5'-ends. Mismatches at the 3'-end of a primer can severely impede PCR DNA synthesis in these methods. For Connection and Inverse PCRs, the mismatches may be placed in the middle or the 5'-end portion of a primer.

In the Megaprimer method, the choice of direction (forward or reverse) of a middle mutagenic primer is determined by the desired position of the mutation in relation to its outside primers. The direction of a mutagenic primer should be chosen to make the megaprimer shorter. To a certain extent, the shorter the mutagenic primer, the more efficient the mutagenesis PCR is.

The optimal positions of two outside primers are often determined by the need of the research project. For expression studies, these positions are usually the initiation and the termination sites of a gene. In some cases, it may also be only part of the coding region of the wild -type gene, provided a pair of unique restriction sites exists or is designed within the wild-type insert.

Usually, once a mutant insert is produced by mutagenesis PCR, e.g., Megaprimer PCR, this insert needs to be cloned into a plasmid or a vector. Thus, a pair of restriction sites should be constructed into the two outsider primers for subsequent cloning. This pair of the restriction sites are usually the same as that of the vector. To ensure a successful digestion, several 5'-terminal base pairs should be designed outside of the restriction site of a primer. The minimal length of such 5'-terminal sequence differs from restriction enzyme to restriction enzyme. This information is available, e.g., in the catalog of New England Biolabs.

The annealing temperature for mutagenesis PCR is typically selected to be 5 °C below the melting temperature (Tm) of the primer. This Tm can be estimated from the primer sequence. An empirical formula, Tm=2 × (number of A and T)+4 × (number of G and C), can be used to estimate this Tm. Programs such as OLIGO (Molecular Biology Insights, Inc.) can also be used to estimate Tm. These programs also help in uncovering any secondary structures that may be present in the primer. If at all possible, secondary structures in primers should be avoided to ensure a successful PCR.

### 3. The use of ddNTP-blocked restriction digestion fragments as an additive to wild-type templates

In order to reduce the number of amplification cycles, and thus to decrease undesired spontaneous mutations, high amounts of plasmid template should be used to obtain sufficient amounts of product. The use of high amounts of template can decrease the mutagenesis efficiency. To retain high efficiencies, ddNTP-blocked restriction digestion fragments can be added to the uncut wild-type template (Ling and Robinson 1995). To prepare this ddNTP-blocked restriction digestion fragments, the wild-type plasmid template is first digested with an appropriate restriction enzyme(s). Digestion with this enzyme or enzymes produces at least one fragment that contains the entire megaprimer region and is shorter than the full-length mutant product. The resulting fragment (non-purified) is then treated with 1 mM each of four dideoxynucleotide phosphates (ddNTPs) for 15 min at 72 °C in the presence of 2 U of Pfu and appropriate buffer, to block the 3'-ends of the restriction digestion fragments. This blockage prevents the fragments from acting as primers which anneal onto templates and form wild- type products. When these fragments were added at 5–50 fmol/100 µl to the uncut templates, the mutagenesis efficiency was increased greatly over a wide range of mutagenic primer and template amounts. We speculate that the ddNTP-blocked digestion fragments serve as a more efficient template than the uncut plasmid, for the megaprimer amplification in the first stage of the One-STEP Megaprimer PCR method.

### 4. A rapid way to check the success of cloning

Because some clones may not contain any inserts or may not contain the desired correct-size inserts, it is helpful to screen clones prior to the preparation and sequencing of the DNAs. Colony PCR is a quick procedure for screening bacterial colonies-transformants containing the DNA clones. Briefly, a colony is picked by a sterile pipette tip, and the tip is then rinsed by pipetting up and down three times into 50 µl LB medium containing the desired antibiotic, such as ampicillin. Then, 1 µl each from five colony solutions are pooled together. One µl from each pool is then boiled for 3 min. This resulting solution is used as a template for a PCR, using two primers which are derived from the vector and flank of the insert. Alternatively, the two outside primers used in the mutagenesis PCR can also be used. Any pool of colonies that gives a correct size product can then be further screened by an ensuing PCR using the same primers to identify individual colonies from a pool. The identified clones are inoculated and analyzed further by preparing, digesting and sequencing the DNA.

## ◼ References

Bennett BL, Molenaar AJ (1994) Cloning of PCR products can be inhibited by Taq polymerase carryover. Biotechniques 16:32, 37

Cline J, Braman JC, Hogrefe HH (1996) PCR fidelity of Pfu DNA polymerase and other thermostable polymerases. Nucleic Acids Res 24:3546–3551

Deng WP, Nickoloff JA (1992) Site-directed mutagenesis of virtually any plasmid by eliminating a unique site. Anal Biochem 200:81–88

Hanahan D (1983) Studies on transformation of Escherichia coli with plasmids. J Mol Biol 166:557–580

Ling M, Robinson B (1995) A One-STEP PCR method of site-directed mutagenesis of large gene-cassettes with high efficiency, yield and fidelity. Anal Biochem 230:167–172

Ling M, Robinson BH (1997) Approaches of DNA mutagenesis: an overview. Anal Biochem 254:157–178

Sambrook J, Fritsch EF, Maniatis T (1989) Molecular cloning: a laboratory manual, 2nd edn. Cold Spring Harbor Laboratory Press, Cold Spring Harbor, New York

Smith M (1985) In vitro mutagenesis. Annu Rev Genet 19:423–462

## ◼ Suppliers

Amersham-Pharmacia Biotech, Bjorkgatan 30, 751 84 Uppsala, Sweden (e-mail: ts-molbio@am.apbiotech.com, Tel.: +46-800-5263593, Fax: (800) FAX-3593, www.apbiotech.com)

Bio 101, Inc., P.O. Box 2284, La Jolla, California 92038–2284, USA (e-mail: technical@bio101.com, Tel.: +1-8004246101 Fax: +1-760-5980116, www.bio101.com)

Clontech Laboratories, Inc., 1020 East meadow Circle, Palo Alto, California 94303–4230, USA (e-mail: tech@clontech.com, Tel.: +1-800-662-clon, Fax: +1-(800-4241350, www.clontech.com)

Life Technologies, (Gibco-BRL Products), Grand Island, New York, USA (e-mail: info@lifetech.com, Tel.: +1-800-8286686, Fax: +1-800-3521468, www.lifetech.com)
Invitrogen, 1600 Faraday Avenue, Carlsbad, California 92008, USA (e-mail: tech × service@invitrogen.com, Tel.: +1-800-9556288 Fax: +1-760-6037201, www.invitrogen.com)

Molecular Biology Insights, Inc. 8685 US Highway 24, Cascade, Colorado 80809–1333, USA (e-mail: rychlik@mbinsights.com, Tel.: +1-800-7474362, Fax: +1-719-6847989)

New England Biolabs, Inc., 32 Tozer Road, Beverly, Massachusetts 01915–5599, USA (e-mail: info@neb.com, Tel.: +1-800-6325227, Fax: +1-978-921–1350, www.neb.com)

Perkin-Elmer Applied Biosystems, 850 Lincoln Center Drive, Foster City, California 94404, USA (Tel.: +1-800-8316844, Fax: +1-650-6385875, www.perkin-elmer.com/ab)

Qiagen, Inc., 28159 Avenue Stanford, Valencia, California 91355, USA (Tel.: +1-800-4268157 Fax.: +1-800-3627737, www.qiagen.com)

Sigma, P.O. Box 14508, St. Louis, Missouri 63178–9916, USA (e-mail: sigma-techserv@sial.com, Tel.: +1-800-5218956, Fax: +1-800-3255052, www.sigma.sial.com)

Stratagene Cloning Systems, 1101 North Torrey Pines Road, La Jolla, California 92037, USA (e-mail: techservices@stratagene.com, Tel.: +1-800-4245444 x3, Fax: +1-619-5350045, www.stratagene.com)

# Long RT-PCR Cloning - Amplification of Full-Length Enterovirus Genomes

KATHLEEN SIMPSON and JOHN W. GOW

## Introduction

At the present time, there is widespread interest in the use of polymerase chain reaction (PCR) amplification techniques to generate DNA or cDNA sequences for target identification, sequence determination, subsequent use as molecular probes, or as functional units which can be expressed in appropriate vector systems. This chapter describes the technique of Long RT-PCR followed by cloning of the PCR products into plasmid vectors. Essentially, this is a method for copying RNA into first-strand cDNA using the enzyme reverse transcriptase (RT) and then PCR amplifying the cDNA to give rise to a double stranded cDNA copy of the original RNA target sequence. The basic technique has been taken to a more advanced stage by the introduction of reverse transcriptase enzymes such as SUPERSCRIPT II RNase H⁻ (Invitrogen Liefe Technologies Ltd) which allow longer stretches of RNA to be copied into DNA and Taq polymerases such as rTth DNA polymerase, XL (PE Applied Biosystems), Deep Vent DNA polymerase (New England Biolabs) and the polymerase mix provided with the Advantage 2 PCR Enzyme System (BD Biosciences, Clonetech, UK) which have high fidelity rates and can generate much longer PCR products. By incorporating restriction enzyme sites at the ends of the amplification primers,

K. Simpson, J.W. Gow (✉) (e-mail: gora20@udcf.gla.ac.uk,
Tel.: +44-141-2012465, Fax: +44-141-2012993)
University of Glasgow Department of Neurology, Southern General Hospital, 1345 Govan Rd., Glasgow G51 4TF, Scotland, UK

Springer Lab Manual
R.C. Bird, B.F. Smith (Eds.) Genetic
Library Construction and Screening
© Springer-Verlag Berlin Heidelberg 2002

it is possible to clone these long PCR cDNA fragments into suitable cloning sites on a large number of commercial vectors.

This chapter also describes the amplification and cloning of enteroviral RNA sequences. These viruses, which are members of the family Picornaviridae, are single-stranded positive sense RNA molecules of approximately 7.4 Kb comprising open coding regions of 6.6 Kb with non-translated regions at both the 5' and 3' termini. The infectious viral RNA has a polyadenylated 3' end and functions as a mRNA template for the synthesis of viral proteins. In essence the viral RNA genomes function as eukaryotic mRNA and so the techniques described below can be applied to any suitable RNA species.

## ▨ Outline

The procedures described below are used for the long RT-PCR amplification and cloning of fragments of RNA several kilobases in length. The initial step is the purification of good quality RNA followed by the synthesis of long complementary DNA (cDNA). The single stranded cDNA is then used as the template for PCR amplification. Amplification may be carried out using oligonucleotide primers which have restriction enzyme sites incorporated at the appropriate ends for subsequent cloning. Alternatively, it is possible to ligate commercial restriction enzyme cartridges onto the cDNA prior to cloning or to carry out blunt end ligations with the PCR products themselves (but be aware that a few base pairs may be missing at the 5' or 3' ends which can affect subsequent expression of the cDNA). PCR products are normally purified by horizontal agarose gel electrophoresis prior to insertion into a suitable plasmid (phage, cosmid or other) vector. The method described here which we have successfully used in our laboratory is outlined in Fig. 1 (see also Gow et al. 1996; Lindberg et al. 1997; Martino et al. 1999; Lindberg et al. 1999).

## Cloning of Full-length Enteroviral cDNA

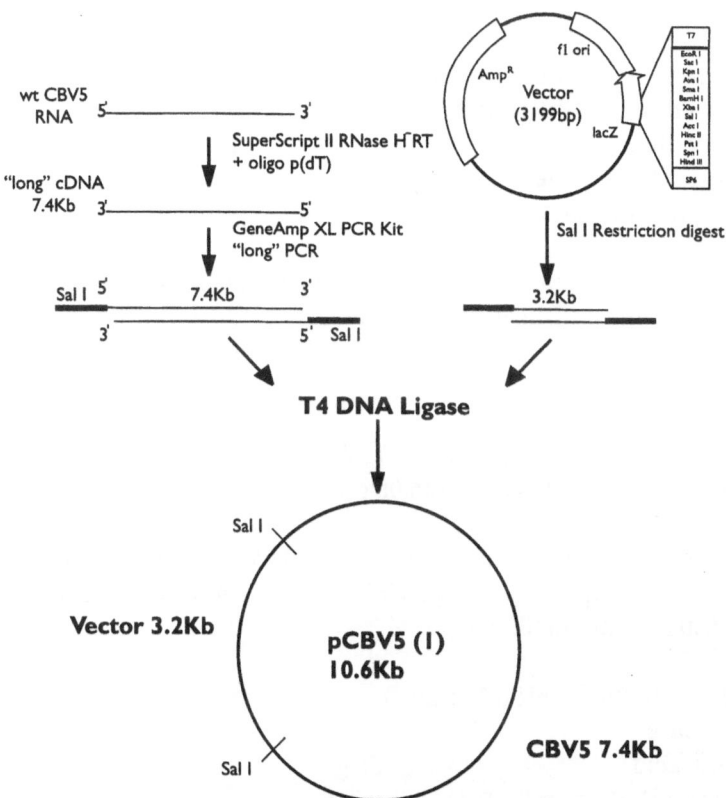

**Fig. 1.** Outline diagram of the Long RT-PCR and cloning procedures. A cDNA copy of CBV5 viral RNA is reverse transcribed, PCR amplified and the full-length amplicons cloned into the Sal I site of the multiple cloning region of the 3.2-Kb vector.

## Materials

**Equipment**    Unless otherwise specified, all centrifugation is carried out in a microcentrifuge at 13,000 rpm.

**Buffers**    Denaturing Solution pH 7.0 (200 ml)

| | |
|---|---|
| guanidinium thiocyanate (sodium salt) | 95 g |
| tri-sodium citrate | 1.47 g |
| N-lauryl sarcosine | 1 g |
| 2-mercaptoethanol | 1.4 ml |

Place weighed reagents in an autoclaved beaker. Add 60 ml sterile $H_2O$ and stir/heat gently to dissolve. Add the 2-mercaptoethanol and adjust final volume to 200 ml. Store at 4 °C, use within 2 months.

2 M Sodium Acetate pH 4.0 (500 ml)

| | |
|---|---|
| sodium acetate.$3H_2O$ | 136.08 g |

Add 40 ml sterile distilled $H_2O$ to 136.08 g of sodium acetate trihydrate and pH to 4.0 with glacial acetic acid. Adjust final volume to 500 ml with distilled $H_2O$.

10 × TBE Buffer pH 8.3 (1 liter)

| | |
|---|---|
| Tris base | 108 g |
| Boric acid | 55 g |
| EDTA disodium salt.$2H_2O$ | 9.3 g |

Place weighed reagents in an autoclaved beaker. Add 800 ml sterile $H_2O$ and stir to dissolve. Adjust final volume to 1 liter with distilled $H_2O$ (pH should be 8.3).

0.5 M EDTA pH 8.0 (1 liter)

| | |
|---|---|
| EDTA disodium salt.$2H_2O$ | 186.18 g |

Weigh out reagent, add 800 ml sterile $H_2O$ and stir vigorously to dissolve. Adjust pH to 8.0 with approximately 20 g sodium hydroxide pellets and adjust final volume to 1 liter. Dispense into 100-ml aliquots and sterilise by autoclaving.

## ▣ Procedure

### Preparation of RNA

RNA for the Long RT-PCR procedure can be prepared by several different methods including total RNA or poly-A+ mRNA. The main requirement for the successful synthesis of long cDNA is, of course, good quality RNA. Commercial kits are available such as the RNAgents Total RNA Isolation System (Promega, UK) which are designed to rapidly isolate high quality RNA for use in applications such as RT-PCR. In practice, we have found that total RNA prepared by a modification of the acid guanidinium phenol chloroform (AGPC) method described by Chomczynski and Sacchi (1987) yields RNA of sufficient quality to prepare cDNA of at least 10 Kb in length. The AGPC method used routinely is as follows:

1. Suspend $1–3 \times 10^7$ tissue culture cells or 3 mm$^3$ tissue in 1 ml of denaturing solution and homogenise in a hand-held Dounce homogeniser (10–20 strokes).

2. Add 100 µl of 2 M sodium acetate pH 4.0 and mix well.

3. Add 1 ml of phenol pH 4.3 and mix. Add 200 µl of chloroform/iso-amyl alcohol (49:1 v/v).

4. Transfer solution to a disposable capped 14-ml polypropylene centrifuge tube and vortex for 10 s.

5. Incubate on ice for 15 min prior to centrifugation at 11,000 rpm (15,000 g) for 20 min, 4 °C (Sorvall SS-34 rotor)

6. Remove the top aqueous phase to a fresh tube and add 1 ml of ice-cold isopropanol. Precipitate on dry ice for 30 min and centrifuge as in step 5.

7. Pour off the supernatant and resuspend the pellet in 300 µl of denaturing solution. Transfer to a 1.5-ml microcentrifuge tube and add 600 µl of ice cold 100 % analar grade ethanol. Precipitate on dry ice for 30 min.

8. Centrifuge at 13,000 rpm for 20 min, 4 °C. Remove supernatant.

9. Add 1 ml of ice cold 70 % ethanol and recentrifuge for 10 min. Remove supernatant.

10. Repeat step 9 and lyophilise RNA pellet. Resuspend in appropriate volume of sterile $H_2O$ (typically 60 µl) and store at –70 °C.

11. The quality of the RNA can be assessed by electrophoresis through a 1 % agarose gel in 1 × TBE buffer. The 28S and 18S RNA bands should be readily visible following ethidium bromide staining (1 µg/ml final concentration) (10 mg/ml stock solution, Sigma) and UV illumination (320 nm).

### Long cDNA synthesis

Long cDNA is prepared from aliquots of total RNA using a 15-mer oligo-p(dT) primer (Roche Diagnostics Ltd) and the enzyme SUPERSCRIPT II RNase H⁻ Reverse Transcriptase (Invitrogen Life Technologies Ltd). Other reverse transcriptase enzymes are commercially available but we have noted that this enzyme can synthesise increased amounts of first strand cDNA and give high yields of full-length products.

1. Add the following components to an RNase, DNase-free microcentrifuge tube:
   RNA (1 µg)
   15-mer oligo p(dT) (0.5 µg)
   sterile $H_2O$ to a volume of 12 µl.

2. Heat to 70 °C for 10 min and cool rapidly on ice. Briefly centrifuge the reaction mix to collect the contents.

3. To each microcentrifuge tube add:
   4 µl 5 × first strand buffer (250 mM Tris-HCL, pH 8.3, 375 mM KCl, 15 mM MgCl)
   2 µl 0.1 M DTT
   1 µl 10 mM dNTP solution (10 mM each of dATP, dTTP, dCTP, dGTP)

4. Incubate at 46 °C for 2 min and add 200 units of SUPERSCRIPT II. Gently mix tube contents and incubate again at 46 °C for 1 h.

5. Stop the reaction by heating at 70 °C for 15 min. cDNA is then used immediately for PCR amplification or stored at –20 °C until required.

The length of the cDNA can be evaluated by several methods. By incorporating a trace of $^{32}$P-dCTP into the cDNA reaction mixture, the length of the cDNA can be visualised after electrophoresis through 1 % agarose gels followed by exposure of the gels to X-ray film. An alternative method is to PCR amplify a short region (100–400 bases)of the cDNA at the 5' end of the cDNA sequence. The cDNA synthesis starts at the 3' end, thus if the short PCR product is targeted at the 5' end, and is successful, it follows that the cDNA synthesis has produced cDNA of the required length. This second method also confirms that the long cDNA is of a sufficient quality to be amplifiable.

## Long RT-PCR amplification

Several kits are now commercially available from a number of major suppliers for the amplification of cDNA. We would recommend that, initially, workers new to the technology should purchase one of the kits and evolve their own modifications of standard protocols to suit specific target sequences and PCR primers and then purchase individual reagents as required. Here, we describe a modification of the GeneAmp XL PCR kit (PE Applied Biosystems) protocol which we use in our laboratory.

Ten percent of the cDNA reaction (2 µl) is used for Long RT-PCR amplification.

1. Add the following reagents to 0.5-ml thin walled PCR microcentrifuge tubes:
   2 µl cDNA
   30 µl GeneAmp 3.3 XL buffer II (supplied by PE Applied Biosystems)
   8 µl 10 mM dNTP mix
   0.5 µg antisense PCR oligo primer
   0.5 µg sense PCR oligo primer
   4.4 µl 25 mM Mg(OAc)$_2$
   4 units rTth DNA polymerase, XL (PE Applied Biosystems)
   1 unit Deep Vent DNA polymerase (New England Biolabs)

2. Overlay samples with 50 µl mineral oil (Sigma) and amplify on a suitable thermocycler. Conditions for the amplification step will depend, of course, on the oligonucleotide primers used as well as the template and the length of the target

**Table 1.** Oligonucleotide PCR Primers

| Primer | Sequence | Polarity |
|--------|----------|----------|
| EntA | 5′– GATC.**GTCGAC**.<u>TTAAAACAGCCTGTGGGTTGTACCCACCCA</u> –3′ | sense |
| EntB | 5′– GATC.**GTCGAC**.<u>TTTTTTTTTTTTTTTTTTTTTTTTTTTCCCGCACC</u><br><u>GAATGCGGATAATTTACCCCTACCGCACCGTTGT</u> – 3′ | antisense |

GATC is a non-coding nucleotide sequence which allows for minor damage at cloning ends.
The sequence highlighted in bold is the Sal I cleavage recognition site.
The underlined sequences are derived from published coxsackie B5 (CBV5) sequence (Zhang et al, 1993). Note that the antisense primer EntB contains a region of poly-T which gives rise to a poly-A tail on the PCR amplicon which is subsequently used for *in vitro* transfection experiments.

sequence. For the oligonucleotide primers shown in Table 1, which amplify a 7.4 Kb target, we use the following:

– an initial denaturation step at 94 °C for 2 min and then 35 cycles of:
– 94 °C for 30 s
– 57 °C for 30 s
– 72 °C for 6 min 30 s

A final elongation step is carried out for 7 min 30 s at 72 °C.

3. Ten percent of the PCR products are analysed by electrophoresis through 1 % agarose gels in 1 × TBE buffer, stained with ethidium bromide and visualised under UV illumination.

### Analysis and purification of PCR product

**Spin-X centrifuge filter unit purification**

1. Long RT-PCR products are electrophoresed through a 1 % agarose gel in 1 × TBE buffer, stained with ethidium bromide and UV visualised (UV exposure time is kept to a minimum to prevent damage to the DNA). The PCR product bands are carefully excised from the gel and placed in a nuclease-free microcentrifuge tube.

2. The agarose slices containing the PCR DNA are freeze/thawed three times and then placed in a Spin-X centrifuge filter unit (0.22 μm cellulose acetate filter; Costar) prior to centrifugation at 13,000 rpm for 20 min at room temperature.

3. The filter from the unit is discarded and the eluate volume left in the microcentrifuge tube is increased to 250 µl with sterile H$_2$O (if necessary).

4. Add an equal volume of phenol/chloroform pH 8.0 (Invitrogen Life Technologies Ltd), vortex for 10 s and centrifuge at 13,000 rpm for 15 min at room temperature. Transfer aqueous phase to a fresh microcentrifuge tube.

5. Repeat step 4.

6. Add 2.5 volumes of 100 % analar grade ethanol and 0.1 volumes 7.5 M ammonium acetate. Precipitate on dry ice for 30 min prior to centrifugation at 13,000 rpm for 20 min at 4 °C.

7. Remove supernatant and wash pellet in 1 ml of 70 % ethanol. Centrifuge at 13,000 rpm for 10 min at 4 °C.

8. Repeat step 7, lyophilise and resuspend pellet in 10 µl sterile H$_2$O.

**Low melting point agarose gel purification**

1. As an alternative to Spin-X column purification, the PCR products can be electrophoresed through 1 % low melting point agarose gels and excised as before.

2. Heat at 70 °C for 20 min followed by 37 °C for 10 min. Add 2–3 volumes of sterile H$_2$O and proceed from step 4 as described above.

**Note:** Several commercial kits are now available (e.g. the QIAquick PCR Purification Kit and the QIAquick Gel Extraction Kit, QIAGEN Ltd.) which are designed to give rapid purification of PCR products. This DNA is then suitable for subsequent applications such as ligation, transformation and transcription.

## Preparation of vector and insert DNA for cloning

There are many different cloning strategies available which can be adapted for use. It is desirable that the PCR primers incorporate restriction enzyme cleavage recognition sites which are present only once in the vector (in the multiple cloning site) and are not present in the PCR target sequence itself. In the example described here, both the vector DNA and the purified Long RT-PCR products are digested with the restriction endonuclease Sal I to generate 5'- cohesive termini. In separate reaction mixes, vector DNA and insert DNA are digested as follows:

1.  To 1 µg of DNA, in a volume of 17 µl of $H_2O$, add 2 µl of 10 × restriction enzyme digestion buffer (SuRE/Cut buffer H, 50 mM Tris-HCl, 100 mM NaCl, 10 mM $MgCl_2$, 1 mM DTE, pH 7.5 at 37 °C; Roche Diagnostics Ltd).

2.  Add 10 units of Sal I restriction endonuclease (Boehringer Mannheim), mix gently and incubate at 37 °C for 90 min.

3.  Inactivate the reaction by adding 0.5 M EDTA pH 8.0 to a final concentration of 10 mM.

4.  Add an equal volume of phenol/chloroform pH 8.0 (Invitrogen Life Technologies Ltd), vortex for 10 s and centrifuge at 13,000 rpm for 15 min at room temperature. Transfer aqueous phase to a fresh microcentrifuge tube.

5.  Repeat step 4.

6.  Add 2.5 volumes of 100 % analar grade ethanol and 0.1 volumes 7.5 M ammonium acetate. Precipitate on dry ice for 30 min prior to centrifugation at 13,000 rpm for 20 min at 4 °C.

7.  Remove supernatant and wash pellet in 1 ml of 70 % ethanol. Centrifuge at 13,000 rpm for 10 min at 4 °C .

8.  Repeat step 7, lyophilise and resuspend pellet in 10 µl sterile $H_2O$. Samples can be stored at –20 °C until required.

### Ligation of vector and insert DNA

The enzyme T4 DNA Ligase is capable of covalently joining two strands of DNA containing 5'-phosphate and 3' hydroxyl termini in a blunt ended or cohesive ended configuration. When cloning a DNA fragment into a DNA vector, the molar ratio of vector to insert DNA will vary depending upon vector type, insert size etc. Ligation reactions usually contain 100–200 ng of vector DNA and a suitable amount of insert which can be calculated from the following equation:

$$\frac{\text{kb size of insert} \times \text{ng of vector}}{\text{kb size of vector}} \times \frac{\text{molar ratio of insert}}{\text{vector}} = \text{ng of insert}$$

1.  A typical 10 µl ligation reaction using a 1:3 vector: insert ratio is as follows:
    100 ng vector DNA (Sal I digested pGEM-3Zf $^+$) (3199 bp)
    740 ng insert DNA (Sal I digested CBV5 long RT-PCR product) (7402 bp)
    1 unit T4 DNA ligase
    1 µl 10 × ligase buffer (300 mM Tris-HCl, pH 7.8, 100 mM MgCl$_2$, 100 mM DTT, 5 mM ATP).

2.  For cohesive ends, incubate the reaction overnight at 15 °C.

3.  Inactivate the reaction by heating to 70 °C for 10 min and store at –20 °C until required.

**Transformation and identification of clones**

In our laboratory we also have used the TOPO XL PCR Cloning Kit (Invitrogen Life Technologies Ltd) which provides a rapid method for the cloning of long PCR products. The protocol does not require the use of ligase or PCR primers designed to incorporate restriction enzyme sites. The PCR vector, pCR-XL-TOPO is provided linearised with single, overhangig 3' deoxythymidine residues. Consequently, PCR products generated from most commercial enzyme mixtures with result in a single deoxyadenosine residue at the 3' end of the amplified fragment ligate efficiently into this vector.

Standard protocols are used for the preparation of competent E. coli, transformation of ligated plasmid DNA into competent cells, identification and restriction enzyme analysis of clones. An example of the full-length RT-PCR amplification of Coxsackie B5 virus (CBV5) is shown in Fig. 2. Lane 1 shows both bands from a Sal I endonuclease digest of a full-length viral clone, as controls, lanes 2 and 3 show the vector and the insert respectively, prior to ligation and transformation.

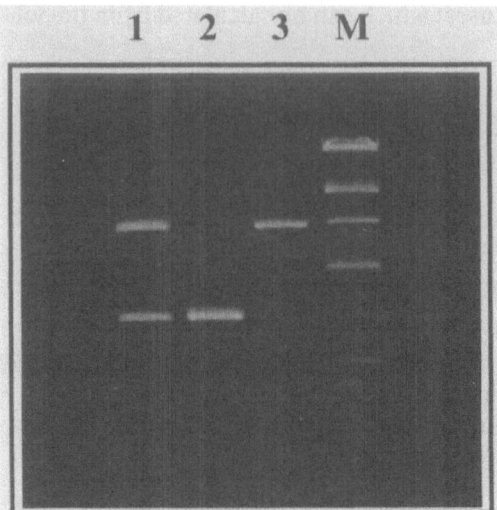

**Fig. 2.** Agarose gel electrophoresis of long RT-PCR amplicon and vector. *Lane 1* Sal I restriction endonuclease digest of cloned full-length CBV5 PCR product in pGEM 3Zf + vector. *Lane 2* Sal I linearised vector (3.2 Kb). *Lane 3* Full-length RT-PCR amplified CBV5 genome (7.4 Kb). *Lane 4* λ Hind III DNA size marker

## Troubleshooting

- For long cDNA synthesis with the reverse transcriptase enzyme it was found that 46 °C was the optimal temperature. With other templates however, different temperatures may be necessary. Try 42 °C as a starting point with increments of 2 °C to a maximum of 50 °C.

- In practice we have found it unnecessary to remove the RNA complementary to the cDNA. However, amplification of some PCR targets may benefit from the removal of the RNA component by incubation with RNase H at 37 °C for 20 min.

# References

Chomczynski P, Sacchi N (1987) Single step method of RNA isolation by acid guanidinium thiocyanate phenol chloroform extraction. Anal Biochem 162:156–159

Gow JW, McGill MM, Behan WMH, Behan PO (1996) Long RT-PCR amplification of full-length enterovirus genome. BioTechniques 20:582–584.

Lindberg AM, Polacek C, Johansson S (1997) Amplification and cloning of complete enterovirus genomes by long distance PCR. J Virol Methods 65:191–199

Lindberg AM, Johansson S, Andersson A (1999) Echovirus 5: infectious transcripts and complete nucleotide sequence from uncloned cDNA. Virus Res 59: 75–87

Martino Ta, Tellier R, Petric M, Irwin DM, Afshar A, Liu PL (1999) The complete consensus sequence of coxsackie B6 and generation of infectious clones by long RT-PCR. Virus Res 64:74–86

Zhang G, Wilsden G, Knowles NJ, McCauley JW (1993) Complete nucleotide sequence of a Coxsackie B5 virus and its relationship to swine vesicular disease virus. J Gen Virol 74:845–853

# Suppliers

Applied Biosystems Division of Perkin-Elmer Ltd., Kelvin Close, Birchwood Science Park North, Warrington, Cheshire, WA3 7PB, UK (Tel.: +44-192-5825650, Fax: +44-192-5282502)

Roche Diagnostics Ltd., Bell Lane, Lewes, East Sussex, BN7 1LG, UK (Tel.: +44-808-100 9998, Fax: +44-808-100 8060)

Life Technologies (Gibco/BRL) Ltd., 3 Fountain Drive, Inchinnan Business Park, Paisley, PA4 9RF, UK (Tel.: +44-800-269210, Fax: +44-800-243 485)

Sigma-Aldrich Co Ltd., Fancy Rd, Poole, Dorset, BH12 4QH, UK (Tel.: +44-800-717181, Fax: +44-800-378785)

Promega UK Ltd., Delta House, Chilworth Research Centre, Southampton, SO16 7NS, UK (Tel.: +44-800-378994, Fax: +44-800-181037)

New England Biolabs Ltd., Knowl Piece, Wilbury Way, Hitchin SG4 0TY, UK (Tel: +44-800-318486, Fax: +44-800-435682)

Costar UK Ltd., 10 The Valley Centre, Gordon Rd,
High Wycombe, Bucks HP13 6EQ, UK (Tel.: +44-494-471207,
Fax: +44-494-464891)

QIAGEN Ltd., Boundary Court Gatwick Rd, Crawley,
West Sussex RH10 2AX, UK (Tel.: +44-1293-422911,
Fax: +44-1293-422922)

BD Biosciences Clonetech, 21 Between Towns Road, Cowley,
Oxford, OX4 3LY, UK (Tel.: +44-1865-781688,
Fax: +44-1865-781627)

# Construction of cDNA Libraries from Small Quantities of Total RNA Using Template Switching Catalyzed by M-MLV Reverse Transcriptase

YORK Y. ZHU, ALEX CHENCHIK, ROGER LI, FLORENCE Y. HSIEH, and PAUL D. SIEBERT

## Introduction

Generation of high quality cDNA libraries with a comprehensive representation of the original mRNA population requires relatively large amounts (5–50 µg) of poly (A)$^+$ RNA which is difficult to obtain when the amount of biological material is limited ( e.g., rare and unstable cell lines, microdissected cancer cells, biopsy materials, pathological specimens, embryonic and neuron tissues, cells in body fluids and so on). To circumvent this problem, several PCR-based technologies for amplification of cDNA from small amounts of total RNA have been described (Froussard 1993; Bertioli et al. 1994; Korneev et al. 1994). Basically, the amplification of a total cDNA population requires that universal primer binding sites are available at both cDNA ends. An arbitrary sequence can easily be imposed at the 5' cDNA end by priming reverse transcription from the poly (A)$^+$RNA fraction of total RNA by oligo (dT) or anchored oligo (dT) primer. Several strategies have been developed to add a determined sequence (anchor) at the 3' end of the first-strand cDNA. These strategies include: (1) oligo (dG) or oligo (dA) tailing by terminal deoxynu-

Y.Y. Zhu, A. Chenchik, R. Li, F.Y. Hsieh, P.D. Siebert
(e-mail: paul_siebert@BD.com, Tel.: +1-650-4248222 ext 1465,
Fax: +1-650-3540776)
Gene Cloning and Analysis Department, BD Biosciences Clontech,
1020 East Meadow Circle, Palo Alto, California 94303-4230, USA

Springer Lab Manual
R.C. Bird, B.F. Smith (Eds.) Genetic
Library Construction and Screening
© Springer-Verlag Berlin Heidelberg 2002

cleotidyltransferase (Bertioli et al. 1994; Korneev et al. 1994); ( 2) the use of T4 RNA ligase to covalently attach a single-stranded (ss) anchor oligonucleotide to the 3' end of the ss cDNA (Apte and Siebert 1993); (3) the ligation of double-stranded (ds) adaptors to both ends of the ds cDNA (Frohman et al. 1988); and (4) the removal of the 7-MeGppp cap structure followed by ligation of an anchor sequence to the 5' end of the decapped mRNA by T4 RNA ligase (Fromont-Racine et al. 1993).

There are two major drawbacks in current PCR-based cDNA library construction methods in comparison with conventional non-PCR-based technologies. The first limitation is the use of multistep protocols where reverse transcription is followed by additional precipitation, extraction and enzymatic steps before the cDNA amplification step. As a result, these technologies require careful optimization of all steps and inevitably result in loss of starting material. The second limitation is the use of PCR technology itself which is biased for more efficient amplification of short, truncated cDNAs rather than amplification of long, full-length cDNAs. As a result, the cDNA libraries contain an under-representation of full-length clones, and in some cases, a loss of particular cDNAs.

Here, we describe a simplified PCR-based cDNA library construction method, which combines in one step, first-strand cDNA synthesis and addition of an arbitrary anchor sequence at the 3' end of the ss cDNA. This method simultaneously employs the two intrinsic properties of Moloney murine leukemia virus reverse transcriptase (M-MLV RT) - reverse transcription of an mRNA template and template switching activity (Kulpa et al. 1997; Zhu et al. 1998). Anchored ss cDNA can then be efficiently amplified by long and accurate PCR (Barnes 1994) with a high yield of full-length ds cDNA products and comprehensive representation of the starting mRNA population.

## Outline

Figure 1 shows a flowchart of this method. Briefly, the first-strand cDNA synthesis is primed from the poly (A)$^+$ RNA fraction of total RNA by a lock-docking oligo (dT) primer (Borson et al. 1992). A synthetic template-switching (TS) oligonucleotide which contains arbitrary oligodeoxynucleotide sequences and three con-

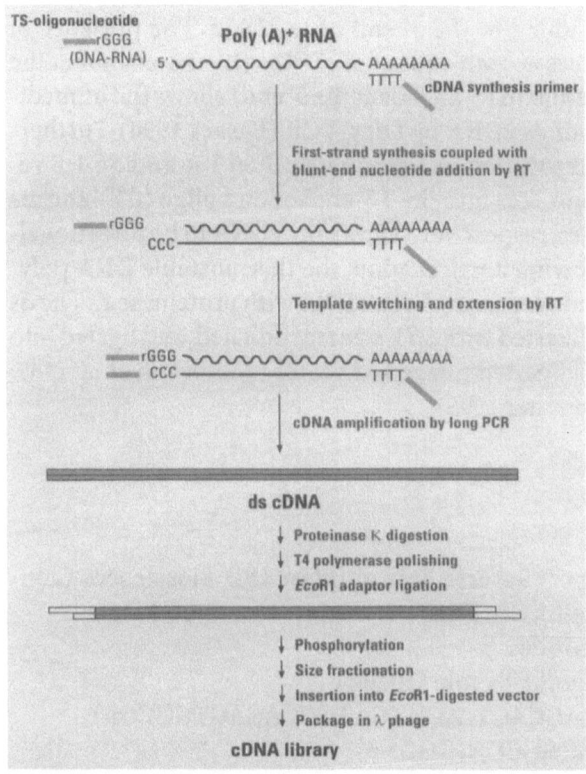

SfiIA                              SfiIB

5'-GGCCATTACGGCC-3'               5'-GGCCGCCTCGGCC-3'

3'-CCGGTAATGCCGG-5'               3'-CCGGCGGAGCCGG-5'

**Fig. 1.** Scheme of the template switching-based cDNA library construction method.

secutive oligoribonucleotide Gs (a r(G)$_3$ stretch) is also included in the first-strand cDNA synthesis mixture. When the MMLV reverse transcriptase (RT) reaches the 5' end of the mRNA, the enzyme's terminal transferase activity adds additional non-template nucleotides (Clark 1988; Kulpa et al. 1997). Based on the analysis of the 3' end structure of these extended products for several model templates, we find that M-MLV RT shows preference for the addition of 2–5 non-template dC nucleotides, resulting in an interaction with the r(G)$_3$ stretch of TS-oligonucleotide, further promoting the RT to switch templates and to continue replicating to the 5' end of the TS-oligonucleotide (Kulpa et al. 1997; Chenchik et al. 1998). As a result, a universal PCR priming site is automatically added to the 3' end of ss-cDNA. The presence of known sequences at both ends of ss cDNAs (the TS-anchor at the 3' end and the oligo (dT)-anchor at the 5' end) allows the immediate generation of ds cDNA by Long-PCR (Barnes 1994). Furthermore, the incorporation of asymmetrical *Sfi* I A and *Sfi* I B restriction enzyme sites into the TS-anchor and oligo (dT)-anchor oligonucleotides, respectively, allows the cDNA to be directionally cloned. Following amplification, the thermostable DNA polymerases are removed from the ds cDNA with proteinase K. The ds cDNA is then digested with *Sfi* I, size fractionated, and ligated into an *Sfi* I (A&B)-digested phagemid vector (Sambrook et al. 1989; Watson and Demmer 1995).

## Materials

- PowerScript™ Reverse Transcriptase (BD Biosciences Clontech, Palo Alto, CA, USA)
- Oligonucleotides
  (1) TS-oligonucleotide (10 µM)
      5'-d(AAGCAGTGGTATCAACGCAGAGTGGCCAT-
      TACGGCC)r(GGG)-3'
  (2) cDNA Synthesis/3' PCR Primer (10 µM)
      5'-ATTCTAGAGGCCGAGGCGGCCGACATG-
      d(T)30N–1N-3'
      N = A, G, C, or T; N–1 = A, G, or C)
  (3) 5' PCR Primer (10 µM)
      5'-AAGCAGTGGTATCAACGCAGAGT-3'
- 5 × first-strand cDNA synthesis buffer
  250 mM Tris-HCl, pH 8.0
  375 mM KCl
  30 mM $MgCl_2$
- dNTP mix (10 mM each of dATP, dCTP, dGTP and dTTP; Pharmacia, Piscataway, NJ, USA)
- Dithiothreitol (DTT, 20 mM)
- Advantage™2 Polymerase mix and buffer (BD Biosciences Clontech, Palo Alto, CA, USA): a mixture of an N-terminal deletion mutant of Taq DNA polymerase, Deep Vent DNA polymerase (New England BioLabs, Beverly, MA, USA), and TaqStart™ antibody (BD Biosciences Clontech, Palo Alto, CA, USA).

- Deionized $H_2O$ (Mili-Q-filtered, not DEPC-treated)
- DNA size markers (1 kb DNA ladder; Life Technologies, Gaithersburg, MD, USA)
- 1 × TAE electrophoresis buffer
  40 mM Tris-acetate, pH 8.3
  1 mM EDTA
- DNA Thermal Cycler. All cycling parameters were optimized by using a Perkin-Elmer GeneAmp PCR Systems 2400/9600 Thermal Cycler (PERKIN-ELMER, Foster City, CA, USA). For different types of thermal cyclers, the cycling parameters must be optimized
- Proteinase K (20 µg/µl, Qiagen, Valencia, CA, USA)
- Sfi I enzyme (20 units/µl, Biolabs, Vancouver, BC, Canada)
- 10X Sfi I buffer (Biolabs, Vancouver, BC, Canada)
- 100X BSA
  20 mM $KPO_4$ (pH 7.0)
  50 mM NaCl
  0.1 mM EDTA
  5 % glycerol
- T4 DNA ligase (400 units/µl, New England Biolabs, Beverly, MA, USA)
- 10 × DNA ligation buffer
  500 mM Tris-HCl (pH 7.8)
  100 mM $MgCl_2$
  100 mM DTT
- ATP (10 mM)
- CHROMA SPIN™ + TE-400 Columns (BD Biosciences Clontech, Palo Alto, CA, USA)
- 1 × Column buffer
  10 mM Tris-HCl (pH 7.4)
  30 mM NaCl
  0.5 mM EDTA
- EDTA (0.2 M)
- Ammonium acetate (4 M)
- Sodium acetate (3 M; pH 4.8)
- Glycogen (20 µg/µl)
- 95 % ethanol
- 80 % ethanol
- 1 % xylene cyanol
- Phenol:chloroform:isoamyl alcohol (25:24:1)
- Chloroform:isoamyl alcohol (24:1)

## ▨ Procedure

### Preparation of RNA

For purification of total RNA preparation we recommend the acid guanidinium thiocyanate-phenol-chloroform method (Chomczynski and Sacchi 1987).

### cDNA synthesis

**First strand cDNA synthesis**

1. Combine the following reagents in a sterile 0.5-ml reaction tube:

| | | |
|---|---|---|
| 1–3 µl | RNA sample (0.05–1 µg total RNA) | |
| 1 µl | cDNA synthesis primer (10 µM) | |
| 1 µl | TS-oligonucleotide (10 µM) | |
| – µl | Deionized H$_2$O | |
| 5 µl | Total volume | |

2. Mix contents and spin the tube briefly in a microcentrifuge.
3. Incubate the tube at 72 °C for 2 min. Cool on ice for 2 min.
4. Spin the tube briefly in a microcentrifuge.
5. Add the following to the reaction tube:

| | |
|---|---|
| 2 µl | 5 × First-strand buffer |
| 1 µl | DTT (20 mM) |
| 1 µl | dNTP(10 mM) |
| 1 µl | Power Script Reverse Transcriptase |
| 10 µl | Total volume |

6. Mix by gentle pipetting and spin the tubes briefly in a microcentrifuge.
7. Incubate the tube at 42 °C for 1 h in an air incubator.
8. Place the tube on ice to terminate first-strand synthesis.

9. If you plan to proceed directly to the PCR step, take a 2 μl aliquot from the first-strand synthesis and place into a clean, prechilled 0.5-ml tube.

10. The remaining first-strand reaction mixture can be stored at −20 °C for up to 3 months.

1. Preheat a hot lid PCR thermal cycler to 95 °C.

**Amplification of the cDNA**

2. Combine the following components in the reaction tube:

| | |
|---|---|
| 2 μl | First-strand cDNA |
| 80 μl | Deionized H$_2$O |
| 10 μl | 10 × Advantage 2 PCR buffer |
| 2 μl | dNTP mix (10 mM each) |
| 2 μl | 5' PCR primer (10 mM) |
| 2 μl | cDNA synthesis / 3' PCR primer |
| 2 μl | 50 × Advantage KlenTaq Polymerase Mix |
| 100 μl | Total/volume |

3. Mix contents by gently flicking. Briefly centrifuge the tube and place it in a preheated (95 °C) thermal cycler.

4. Commence thermal cycling using the following program with a hot-lid thermal cycler:
   - Heat the tube at 95 °C for 20 s
   - Then commence* cycles of PCR:
     95 °C      5 s
     68 °C      6 min
     * Refer to the following table to determine the optimal number of cycles to use:

| Total RNA (μg) | Number of cycles |
|---|---|
| 1.0–2.0 | 18–20 |
| 0.5–1.0 | 20–22 |
| 0.25–0.5 | 22–24 |
| 0.05–0.25 | 24–26 |

5. When the cycling is completed, analyze a 5 µl sample of each test tube alongside 0.1 µg of 1 kb DNA size markers on a 1.2 % agarose/ethidium bromide (EtBr) gel in 1 × TAE buffer. The ds cDNA should appar as a 0.4–4 Kb smear. Some distinct bonds may be visible.

6. Proceed to the next step or store ds cDNA at –20 °C until use.

**Proteinase K digestion**

1. In a sterile 0.5-ml tube, pipet 50 µl of amplified ds cDNA (2–3 µg), and add 2 µl of proteinase K (20 µg/µl). Store the remaining ds cDNA at –20 °C (up to 3 months).
   **Note:** Proteinase K treatment is necessary to inactivate the DNA polymerase activity. This protocol is optimized for 2–3 µg (~50 µl/vol) of PCR cDNA product for subsequent cloning and library construction procedures. Too much ds cDNA (>3–4 µg) will yield a low titer library.

2. Mix contents and spin the tube briefly.

3. Incubate at 45 °C for 20 min. Spin the tube briefly.

4. Add 50 µl of deionized $H_2O$ to the tube.

5. Add 100 µl of phenol:chloroform:isoamyl alcohol and mix by continuous gentle inversion for 1–2 min.

6. Centrifuge at 14,000 rpm for 5 min to separate the phases.

7. Move the top (aqueous) layer to a clean 0.5-ml tube. Discard the interface and lower layers.

8. Add 100 µl of chloroform:isoamyl alcohol to the aqueous layer. Mix by continuous gentle inversion for 1–2 min.

9. Centrifuge at 14,000 rpm for 5 min to separate the phases.

10. Move the top (aqueous) layer to a clean 0.5-ml tube. Discard the interface and lower layers.

11. Add 10 µl of 3 M sodium acetate, 1.3 µl of glycogen (20 µg/µl) and 260 µl of room-temperature 95 % ethanol. Immediately centrifuge at 14,000 rpm for 20 min at room temperature.
    **Note:** Do not chill the tube at –20 °C or on ice before centrifuging. Chilling the sample will result in coprecipitation of impurities.

12. Carefully remove the supernatant with a pipette. Do not disturb the pellet.

13. Wash pellet with 100 µl of 80 % ethanol.

14. Air dry the pellet (~10 min) to evaporate off residual ethanol.

15. Add 79 µl of Deionized $H_2O$ to resuspend the pellet.

1.  Combine the following components in a fresh 0.5-ml tube:

| | |
|---|---|
| 79 µl | cDNA |
| 10 µl | 10X *Sfi* Buffer |
| 10 µl | *Sfi* I Enzyme |
| 1 µl | 100X BSA |
| 100 µl | Total volume |

2.  Mix well. Incubate the tube at 50°C for 2 hr.

3.  Add 2 µl of 1 % xylene cyanol dye to the tube above. Mix well.

1.  Label sixteen 1.5-ml tubes and arrange them in a rack in order.

2.  Prepare the CHROMA SPIN-400 Column for drip procedure:
    - Invert the CHROMA SPIN column several times to completely resuspend the gel matrix.
    - Remove the top cap from the column. Use a 1000-µl pipettor to resuspend the matrix gently; avoid generating air bubbles. Then remove the bottom cap and let the buffer drip naturally. (If the column does not drain after 3 min, recap the top cap. This pressure should cause the column to drain).
    - Attach the column to a ring stand.
    - Let the storage buffer drain through the column by gravity flow until you can see the surface of the gel beads in the column matrix. The top of the column matrix should be at the 1.0-ml mark on the wall of the column. If your column contains significantly less matrix, adjust the vol-

ume of the matrix to the 1.0-ml mark using matrix from another column.

- The flow rate should be approximately 1 drop/40–60 sec. The volume of 1 drop should be approximately 40 µl. If the flow rate is too slow (i.e., more than 1 drop/100 sec) and the volume of one drop is too small (i.e., less than 25 µl), you should resuspend the matrix completely and repeat the drip procedure until it reaches the above parameters.

3. When the storage buffer stops dripping out, carefully and gently (along the column inner wall) add 700 µl of column buffer to the top of the column and allow it to drain out.

4. When this buffer stops dripping (~15–20 min), carefully and evenly apply ~100 µl mixture of *Sfi* I-digested cDNA and xylene cyanol dye to the top-center surface of the matrix. An unsmooth matrix surface does not hurt the following fractionation process.

5. Before proceeding to the next step, allow the sample to be fully absorbed into the surface of the matrix (i.e., there should be no liquid remaining above the surface).

6. With 100 µl of column buffer, wash the tube that contained the cDNA, and gently apply this material to the surface of the matrix.

7. Allow the buffer to drain out of the column until there is no liquid left above the resin. When the dripping has ceased, proceed to the next step. At this point, the dye layer should be several mm into the column.

8. Place the rack containing the collection tubes under the column, so that the first tube is directly under the column outlet.

9. Add 600 ml of column buffer and immediately begin collecting single-drop fractions (approximately 35 ml per tube) in tubes #1–16. Cap each tube after each fraction is collected. Recap the column after fraction #16 has been collected.

10. Check the profile of the fractions before proceeding with the experiment. On a 1.1 % agarose/EtBr gel, electrophorese

3 ml of each fraction (separately) in adjacent wells, alongside 0.1 mg of a 1-kb DNA size marker. Run the gel at 150 V for 10 min. (Running the gel longer will make it difficult to see the cDNA bands). Determine the peak fractions by visualizing the intensity of the bands under UV. Collect the first three fractions containing cDNA (in most cases, the fourth fraction containing cDNA is usable. Make sure the fourth fraction matches your desired size distribution). Pool the above fractions in a clean 1.5-ml tube.

11.  Add the following reagents to the tube with 3–4 pooled fractions containing the cDNA: (105–140 µl, respectively)

| | |
|---|---|
| 1/10 vol. | sodium acetate (3 M; pH 4.8) |
| 1.3 µl | glycogen (20 mg/ml) |
| 2.5 vol. | 95 % ethanol (–20°C) |

12.  Mix by gently rocking the tube back and forth.

13.  Place the tube in –20 °C or a dry-ice/ethanol bath for 1 hr. (Optional: you may incubate at –20 °C overnight, which may result in better recovery).

14.  Centrifuge the tube at 14,000 rpm for 20 min at room temperature.

15.  Carefully remove the supernatant with a pipette. Do not disturb the pellet.

16.  Briefly centrifuge the tube to bring all remaining liquid to the bottom.

17.  Carefully remove all liquid and allow the pellet to air dry for ~10 min.

18.  Resuspend the pellet in 7 ml of Deionized H2O and mix gently. The *Sfi* I-digested cDNA is now ready to be ligated into an *Sfi* I(A&B)-digested, dephosphorylated phagemid vector.

**Ligation of cDNA to Sfi I (A & B) digested, dephosphory-lated phage arm**

1.  Label three 0.5-ml tubes and add the indicated reagents as following table:

| Component | 1st ligation (µl) | 2nd ligation (µl) | 3rd ligation (µl) |
|---|---|---|---|
| cDNA | 0.5 | 1.0 | 1.5 |
| λ arm (500 ng/µl) | 1.0 | 1.0 | 1.0 |
| 10 × ligation buffer | 0.5 | 0.5 | 0.5 |
| ATP (10 mM) | 0.5 | 0.5 | 0.5 |
| T4 DNA ligase (400 unit/µl) | 0.5 | 0.5 | 0.5 |
| Deionized $H_2O$ | 2.0 | 1.5 | 1.0 |
| Total Volume (µl) | 5.0 | 5.0 | 5.0 |

Mix the reagents gently. Spin tubes briefly to bring contents to the bottom of the tube.

2.  Incubate tubes at 16 °C overnight.

3.  Perform a separate λ phage packaging reaction for each of the ligations. (Note: λ phage packaging extracts are commercially available. Choose a packaging system that will give you at least $1 \times 10^9$ pfu/µg of DNA. Follow the supplier's recommended protocol and perform a parallel packaging reaction with the control DNA provided in the packaging kit.)

4.  Plating of recombinant phage (Sambrook et al. 1989; Watson and Demmer 1995).

5.  Titer each of the resulting libraries. From the three ligations combined, you should obtain $1-2 \times 10^6$ independent clones. The unamplified libraries can be stored at 4 °C for 2 weeks.

6.  [Optional] If you obtained $<1-2 \times 10^6$ independent clones, you may wish to perform another ligation with the remaining cDNA. For a repeat ligation, use the ratio of cDNA to vector that gave the best results of the initial three ligations. Scale-up the volumes of all reagents according to the amount of cDNA used. Then package and titer this scaled-up ligation.

7.  Amplification of library (Sambrook et al. 1989; Watson and Demmer 1995). The amplified library can be stored at 4 °C for 6–7 months or at –70 °C (in 7 % DMSO) for at least 1 year.

# Results

### Analysis of PCR cDNA products

To assess the efficiency of the approach, we compared the size distribution of the ds cDNA products synthesized from total or poly (A)+ RNA by the conventional (non-PCR) Gubler and Hoffman method with ds cDNA products amplified by the template switching-based method (Gubler and Hoffman 1983). As shown in Fig. 2, when the conventional method was used, the profile of ds cDNA generated from poly (A)+ RNA (Fig. 2B, lane 1) resembled the banding pattern of human skeletal muscle poly (A)+ RNA template (Fig. 2A, lane 1). In contrast, ds cDNA synthesized by the conventional method from total RNA did not match this profile. This material corresponds to cDNA synthesized from the rRNA (Fig. 2B, lane 2). When the template switching method was used, the banding patterns of ds cDNA amplified from both poly (A)+ RNA and total RNA were the same and both resembled the banding pattern of human skeletal muscle mRNA, even when only 50 ng of total RNA was used (Fig. 2C, lanes 1–3). These data demonstrate that cDNA amplification based on template switching allows selective amplification of the poly (A)+ RNA fraction of total RNA. Although the maximum size of ds cDNA was slightly less than the size of ds cDNA generated by conventional cDNA synthesis, the proposed technology still allows the generation of cDNA products up to 6 kb (Fig. 2C).

To evaluate the sensitivity of this method, different quantities of total RNA were used as starting material for cDNA synthesis and amplification. As shown in Fig. 3 (lanes 1–4), size distribution and pattern of the bands in the ds cDNA amplified from 1 µg and 1 ng of total RNA are practically indistinguishable from each other. When the starting concentration of total RNA template was less than 1 ng, the amplification started to be unreproducible, generating shorter amplification products with more distinct bands (Fig. 3). In addition, ds cDNAs amplified from different quantities of total RNA were tested by PCR with two gene-specific primers for the presence of the following genes: interferon-β1 and interferon-γ (low-abundance mRNAs); and glyceraldehyde 3-phosphate dehydrogenase (G3PDH) and β-myosin (high abundance mRNAs). As shown in Fig. 3, all these gene sequences were detected in cDNA amplified from a mini-

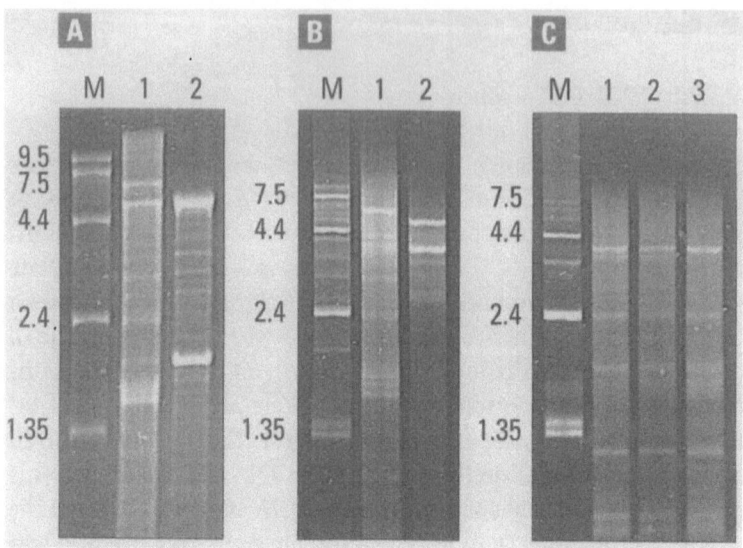

**Fig. 2.** Comparison of gel patterns among RNA, conventional cDNA and cDNA amplified by the template switching-based method: **A** Analysis of gel patterns for RNA was carried out on a formaldehyde agarose/EtBr gel. *Lane M* RNA ladder (Life Technologies, Gaithersburg, MD,USA). *Lane 1* Human skeletal muscle poly (A)+ RNA (500 ng). *Lane 2* Human skeletal muscle total RNA (500 ng). **B** ds cDNA was synthesized from the indicated amount of human skeletal muscle poly (A)+ RNA or total RNA using the conventional Gubler and Hoffman method. *Lane M* ds cDNA synthesized from 500 ng of the RNA ladder. *Lane 1* From 500 ng human skeletal muscle poly (A)+ RNA. *Lane 2* From 500 ng skeletal muscle total RNA. ds cDNA samples were analyzed on a 1.1 % agarose/EtBr gel. **C** ss cDNA was synthesized from the indicated amounts of human skeletal muscle poly (A)+ RNA and total RNA using the template-switching method. ds cDNAs were amplified from 10 % of the corresponding ss cDNAs by PCR. *Lane M* cDNAs generated from 500 ng of the RNA ladder. *Lane 1* From 500 ng of human skeletal muscle poly (A)+ RNA. *Lane 2* From 500 ng of human skeletal muscle total RNA. *Lane 3* From 50 ng of human skeletal muscle total RNA. Five µl of each of the above PCR products were analyzed on a 1.1 % of agarose/EtBr gel

mum of 1 ng of total RNA. Using smaller amounts of starting total RNA gives unreproducible amplification of low-abundance mRNAs. These data demonstrate that a representative cDNA population can be generated by template switching from as little as 1 ng of total RNA. As also shown in previous reports (Belyavsky et al. 1989; Korneev et al. 1994), we find that in order to obtain reliable amplification of very low-abundance mRNAs (single molecule per cell) and achieve a high level of mRNA rep-

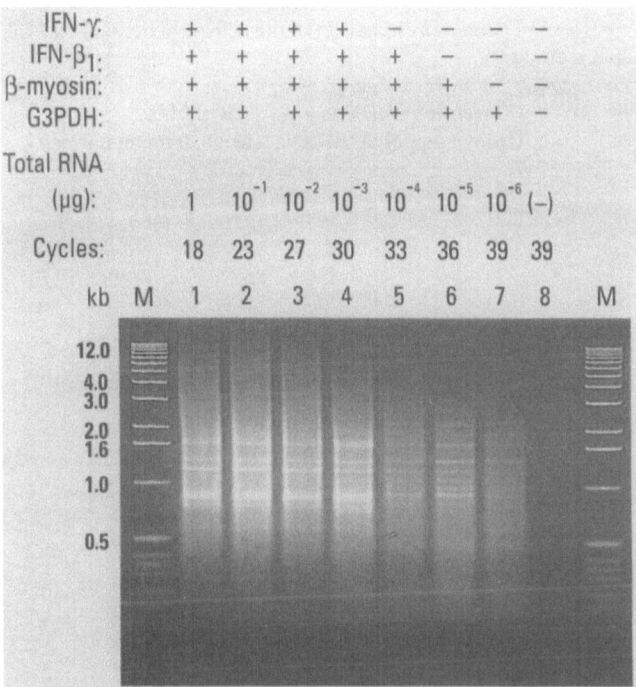

**Fig. 3.** Relationship between gene representation and starting amounts of total RNA. Gene representation was determined by RT-PCR with four human genes: interferon-$\gamma$ (IFN-$\gamma$); interferon-$\beta$1 (IFN-$\beta$1); $\beta$-myosin and glyceraldehyde 3-phosphate dehydrogenase (G3PDH). (+) and (-) indicate the presence and absence of positive PCR signals, respectively. Human skeletal muscle total RNA was used as starting material for first-strand cDNA synthesis with the indicated amounts of starting material (*lanes 1–7*). *Lane 8* Negative control (no RNA). cDNAs were amplified using the two-step PCR program described in Materials and Methods for the indicated number of cycles. Five µl of the corresponding final PCR products were loaded on a 1.1 % agarose /EtBr gel, along with 1 kb DNA size markers (*lane M*)

resentation, the number of PCR cycles should be no more than 25. Based on these data (see also Fig. 3) we suggest using more than 50 ng of total RNA to obtain the most robust amplification.

To further examine the representation of individual mRNAs in cDNA obtained from 50 ng of total RNA, we used PCR to examine the presence of 45 randomly selected genes (cytokines, receptors, oncogenes and housekeeping genes) which belong to different abundance classes (Table 1). As a control we used cDNA synthesized from 5 µg of poly (A)$^+$ RNA by conventional non-PCR based technology. In both cDNA populations we detected

**Table 1.** Comparison of gene representation in the cDNAs generated by PCR vs. conventional method

| Human gene | Conventional cDNAs (from 5 µg polyA+RNA) | PCR cDNAs (from 50 ng total RNA) |
| --- | --- | --- |
| β-Actin | +[a] | + |
| β-Myosin | + | + |
| $\beta_2$ M | + | + |
| c-fms | + | + |
| c-fos | + | + |
| c-jun | + | + |
| c-kit | + | + |
| c-myc | + | + |
| EGF | + | + |
| EGFR-3 | + | + |
| G3PDH | + | + |
| G-CSF | + | + |
| GM-CSF | − | + |
| IGF II | + | + |
| IGFR-1 | + | + |
| IFN-α1 | + | + |
| IFN-β1 | + | + |
| IFN-γ | + | + |
| IL-1 α | + | + |
| IL-1 β | + | + |
| IL-2 | + | + |
| IL-3 | − | + |
| IL-4 | + | + |
| IL-5 | + | + |

the presence of 42 of 45 (93 %) of the genes. Although we cannot exclude the possibility that the amplification procedure can sometimes change the relative concentration of particular cDNAs (for example cDNAs with G/C-rich sequences), these data demonstrate that gene representation in the amplified cDNA is similar to that of cDNA synthesized by the conventional method.

### Quality of template switching-based cDNA library

Based on the strategy shown in Fig. 1, we constructed a human skeletal muscle cDNA library in a λgt11 vector starting with 100 ng of total RNA. As a control, a conventional cDNA library

**Table 1** (*Continued*)

| Human gene | Conventional cDNAs (from 5 µg polyA+RNA) | PCR cDNAs (from 50 ng total RNA) |
|---|---|---|
| IL-6 | + | + |
| IL-7 | + | + |
| IL-8 | + | + |
| IL-10 | + | + |
| IL-11 | + | – |
| IL-2R a | + | + |
| IL-2Rb | + | – |
| IL-4R | + | + |
| IL-6R | + | + |
| NF-kb | + | + |
| p53 | + | + |
| PDGF-A | + | + |
| TFR | + | + |
| TGF- α | + | + |
| TGF-β1 | + | + |
| TGF-β2 | + | + |
| TNF-α | + | + |
| TNF-β | + | + |
| TNF-R1 | + | + |
| TNF-R2 | – | – |
| Positive | 93.3 %(42/45) | 93.3 %(42/45) |

[a] + and – indicate the presence and absence of positive PCR signals, respectively

was constructed using the Gubler and Hoffman method in the same vector starting with 10 µg of poly (A)+ RNA. Both cDNA libraries contained $3 \times 10^6$ independent clones. Insert length analysis of 60 randomly selected clones showed that 95 % of the clones from both cDNA libraries contained inserts within the size range 0.5–6 kb with an average insert of approximately 2 kb. These data suggest that the use of long-distance polymerase chain reaction (LD PCR) lessens the bias in amplification and cloning of short cDNA products, which is a common problem with cDNA library construction technologies utilizing conventional PCR (Akowitz and Manuelidis 1989; Belyavsky et al. 1989; Korneev et al. 1994).

To evaluate the efficiency of full-length cDNA cloning for rather long mRNAs, we screened approximately 10,000 clones from each library with hybridization probes corresponding to the 5' and 3'-ends of the 5 kb human transferrin receptor (TFR) and 6 kb β-myosin mRNAs. The percent of full-length clones was then estimated as the ratio between number of clones which showed signals with 5' and 3' end probes. The percent of full-length clones of β-myosin and TFR was found to be 36 % (5'/3'=52/145) and 46 % (5'/3'=56/121) respectively in the template switching-based library, and 7 % (5'/3'=10/136) and 4 % (5'/3'=7/183) respectively, in the conventional library. These data demonstrate that although the abundance level of β-myosin and TFR are approximately equal in both libraries, the percent of full-length clones is significantly higher in the template switching-based library. In order to explain these unexpected results we speculate that the template switching reaction is most efficient when reverse transcriptase reaches the 5' end of mRNA template. If reverse transcriptase prematurely terminates cDNA synthesis in the regions of RNA with strong secondary structure, the truncated cDNA fragment-RNA hybrid may not be efficiently used by reverse transcriptase for adding additional non-template nucleotides required for template switching. As a result, the amplified cDNA will be enriched for full-length clones. However, short non full-length poly(A)$^+$ RNA fragments could also be efficiently reverse transcribed to the 5' end, followed by the template switching reaction and amplification of non full-length cDNAs by PCR. Clearly, care must be taken to ensure that the RNA is not degraded.

To further characterize the quality of the cDNA library constructed by the template switching-based method, we partially sequenced cDNA inserts from 50 randomly selected clones. The sequences were then analyzed for homology in the GenBank and EMBL data bases. As shown in Table 2, 28 % of the inserts corresponded to well-characterized known genes, 32 % of inserts had homology with expressed sequence tags (ESTs) and 30 % of clones showed no significant homology with entries in the databases and may represent novel genes. The frequency of undesirable sequences such as Alu-like repetitive sequences (4 %), mitochondrial DNA (4 %), ribosomal RNA (0 %) and adaptor dimmers (2 %) were significantly lower, or equal to, the frequency of these sequences in different conventional or PCR-based cDNA libraries (Korneev et al. 1994; Orr et al. 1994; Sudo et

**Table 2.** Summary of a database search[a] of 50 sequences from one-pass sequencing of human skeletal muscle RTTS-PCR cDNA clones

| Category | Frequency Number | (%)[b] |
|---|---|---|
| Known genes | 14 | 28 |
| ESTs | 16 | 32 |
| Novel genes | 15 | 30 |
| Alu-like repetitive sequences | 2 | 4 |
| Mitochondrial DNA sequences | 2 | 4 |
| Ribosomal RNA sequences | 0 | 0 |
| Adaptor dimer | 1 | 2 |
| Total | 50 | 100 |

[a] BLAST (nucleic acid) analysis program (Altschul et al. 1990).
[b] Percent of 50.

al. 1994). These data clearly indicate that at least 90 % of the clones in the cDNA library were derived from the poly (A)$^+$ RNA fraction of the total RNA.

The main innovation of the present work is the use of a template switching reaction to simplify the protocol for construction cDNA libraries using PCR technology. The procedure allows the construction of representative and nearly full-length cDNA libraries from as little as 10–100 ng of total RNA. The quality of these libraries based on results of random sequence analysis achieves the standards of high quality conventional cDNA libraries generated from 5–10 µg of poly (A)$^+$ RNA.

## Troubleshooting

- **No PCR product**
  One or more essential reagents may have been inadvertently omitted from the first-strand synthesis or the PCR reaction. Repeat both of these steps, being careful to check off every item as you add it to the reaction.

- **Size distribution of PCR product less than 3 kb**
  RNA starting material may be degraded, very impure, or too dilute. If you have not already done so, check the quality and

quantity of your RNA by running a sample on a gel. If the RNA seems too dilute, but otherwise of good quality, restart the experiment using more RNA. If the RNA seems degraded, restart the experiment using a fresh lot or preparation of RNA. Also, check the stability of your RNA by incubating a small sample at 37 °C for 2 h, and then running it on a gel parallel to a fresh (unincubated) sample. If the RNA appears to be very unstable by this test, it will not yield good results in the first-strand synthesis.

- **Low yield of PCR product**
  1. Too few thermal cycles used in the PCR step (i.e., PCR undercycling). Another indication of PCR undercycling is a cDNA size distribution <4 kb – if the mRNA source was mammalian. (For some sources, such as many insect species, the normal mRNA size distribution may be <2–3 kb.) If you suspect undercycling, incubate the PCR reaction mixture for two more cycles and recheck the product. If you already used the maximum recommended number of cycles indicated in the Procedure, increase by three more cycles. If increasing the number of cycles does not improve the yield of PCR product, repeat the PCR de novo using a fresh 2 µl aliquot of the first-strand product.
  2. If you still obtain a low yield of PCR product, it may be due to a low yield of first-strand cDNA. Possible problems with the first-strand reaction include: a mistake in the procedure (such as using a suboptimal incubation temperature or omitting a component) or not using enough RNA in the reaction. It is also possible that the RNA has been partially degraded (by contaminating RNases) before or during the first-strand synthesis. Reminder: problems with the first-strand cDNA synthesis can be more easily diagnosed if you perform parallel reactions using the control RNA provided in the kit. If good results were obtained with the control RNA but not with your experimental RNA, then there may be a problem with your RNA.

The easiest way to check the quality of the first-strand cDNA is by using a small sample of it as a PCR template with 3' and 5' housekeeping gene-specific primers, such as human β-actin. If the first-strand synthesis has been successful, a PCR product of the expected size will be generated.

- **No bright bands distinguishable in the PCR product**
  As explained in Figs. 2 and 3, for most mammalian RNA sources, there should be several bright bands distinguishable against the background smear when a sample of the PCR product is run on a gel. If bright bands are expected but are not visible, and the background smear is very intense, this is indicative of PCR overcyling. If you suspect overcycling, then the PCR step must be repeated de novo, with a fresh 2 µl sample of first-strand cDNA, using 2–3 fewer cycles.

- **Presence of low-molecular-weight (<0.1 kb) material in the PCR product**
  The raw cDNA (e.g., before size fractionation) is expected to contain some low-molecular-weight DNA contaminants, including unincorporated primers, unligated adaptors, and very short PCR products. However, these small fragments are generally removed from the ds cDNA preparation in the size fractionation step. Note that a preponderance of low-molecular-weight (<0.1 kb) material in the raw PCR product may be indicative of overcycling. If you suspect overcycling, then the PCR must be repeated de novo with a fresh 2 µl sample of first-strand cDNA, using 2–3 fewer cycles.

- **Presence of low-molecular-weight (<0.1 kb) material in the size-fractionated cDNA**
  The size fractionation columns and the procedure have been optimized to efficiently remove low-molecular-weight cDNA fragments, small DNA contaminants, and unincorporated nucleotides from the cDNA. Failure to remove low-molecular-weight contaminants will result in a library having a preponderance of very small inserts and/or apparently nonrecombinant clones. To avoid this, be sure to check your column fractions on an agarose gel as described in the Procedure and pool only the two peak fractions containing the most intense cDNA bands, along with the two preceding and two succeeding fractions (e.g., fractions 6–11).

- **Low titer of unamplified library**
  Check to make sure you are using a λ phage packaging system that, according to the manufacturer's specifications, is expected to yield at least $1 \times 10^9$ pfu/µg of control λ DNA.

If the unamplified library has a titer $<10^6$ pfu/µg DNA (all three ligations combined) but the control packaging reaction (using ligated, nonrecombinant $\lambda$ DNA) yielded at least $5 \times 10^8$ pfu/µg, there may be a problem with the ligation of vector to insert. The ligation reaction cannot be checked retroactively because the entire ligation mixture is generally used in the packaging reaction. However, before you repeat the ligations, check the concentration of the size-fractionated cDNA by electrophoresing 1 µl of the resuspended cDNA from Steps 6–19 on an EtBr-containing agarose gel next to a known amount of control DNA and view the gel under UV light. Alternatively, spot 1 µl of the cDNA on an EtBr-containing agarose plate next to small spots of known amounts (10–1000 ng) of control DNA. The concentration of the resuspended cDNA should be in the range of 100–200 ng/µl.

If these control results are within expected parameters, repeat the ligation with your cDNA, adjusting the ratio of cDNA to vector in accordance with which ratio in Table 2 gave the best results in the first attempt. For example, if you obtained the greatest number of plaques using a ratio of 1.5:1.0, then when you repeat the ligations, use ratios of 1.5:1.0, 1.75:1.0, and 2.0:1.0.

- **Low (<75%) recombination efficiency**
  For well digested and dephosphorylated $\lambda$ arms, the background due to religation of nonrecombinant vector molecules should be minimal. This can be checked by performing a control ligation using the vector alone (no cDNA). A high titer, combined with an apparently low recombination efficiency, may indicate that your cDNA population contains a lot of small DNA fragments, such as adaptor dimers, which are preferentially ligated into the vector. If you suspect this may be the case, repeat the cDNA synthesis and check the size distribution of the fractionated cDNA.

- **Small insert sizes**
  If more than 50 % of your clones appear to have very small insert sizes (i.e., <0.4 kb), it is likely that your cDNA preparation has not been successfully size-fractionated. Small cDNA fragments, adaptor-primers, unincorporated primers, and primer-dimers are preferentially ligated to the vector, and thus should be completely removed before the vector ligation

step. Because this problem cannot be retroactively solved, you will have to repeat the cDNA synthesis procedure.

## References

Akowitz A, Manuelidis L (1989) A novel cDNA/PCR strategy for efficient cloning of small amounts of undefined RNA. Gene 81:295–360

Altschul SF, Gish W, Miller W, Myers EW, Lipman DJ (1990) Basic local alignment search tool. J Mol Biol 215:403–410

Apte AN, Siebert PD (1993) Anchor-ligated cDNA libraries: a technique for generating a cDNA library for the immediate cloning of the 5' ends of mRNAs. Biotechniques 15:890–893

Barnes WM (1994) PCR amplification of up to 35-kb DNA with high fidelity and high yield from lambda bacteriophage templates. Proc Natl Acad Sci USA 91:2216–2220

Belyavsky A, Vinogradova T, Rajewsky K (1989) PCR-based cDNA library construction: general cDNA libraries at the level of a few cells. Nucleic Acids Res 17:2919–2932

Bertioli DJ, Smoker M, Brown AC, Jones MG, Burrows PR (1994) A method based on PCR for the construction of cDNA libraries and probes from small amounts of tissue. BioTechniques 16:1054–1058

Borson ND, Salo WL, Drewes LR (1992) A lock-docking oligo(dT) primer for 5' and 3' RACE PCR. PCR Methods Appl 2:144–148

Chomczynski P, Sacchi N (1987) Single-step method of RNA isolation by acid guanidinium thiocyanate-phenol-chloroform extraction. Anal Biochem 162:156–159

Clark JM (1988) Novel non-templated nucleotide addition reactions catalyzed by procaryotic and eucaryotic DNA polymerases. Nucleic Acids Res16:9677–9686

Frohman MA, Dush MK, Martin GR (1988) Rapid production of full-length cDNAs from rare transcripts: amplification using a single gene-specific oligonucleotide primer. Proc Natl Acad Sci USA 85:8998–9002

Fromont-Racine M, Bertrand E, Pictet R, Grange T (1993) A highly sensitive method for mapping the 5" termini of mRNAs. Nucleic Acids Res 21:1683–1684

Froussard P (1993) rPCR: a powerful tool for random amplification of whole RNA sequences. PCR Methods Appl 2:185–190

Gubler U, Hoffman BJ (1983) A simple and very efficient method for generating cDNA libraries. Gene 25:263–269

Korneev S, Blackshaw S, Davies JA (1994) cDNA libraries from a few neural cells. Prog Neurobiol 42:339–346

Kulpa D, Topping R, Telesnitsky A (1997) Determination of the site of first strand transfer during Moloney murine leukemia virus reverse transcription and identification of strand transfer-associated reverse transcriptase errors. EMBO J 16:856–865

Orr SL, Hughes TP, Sawyers CL, Kato RM, Quan SG, Williams SP, Witte ON, Hood L (1994) Isolation of unknown genes from human bone marrow by differential screening and single-pass cDNA sequence determination. Proc Natl Acad Sci USA 91:11869–11873

Sambrook J, Fritsch EF, Maniatis T (1989) Molecular cloning: a laboratory manual, 2nd edn. Cold Spring Harbor Laboratory Press, Cold Spring Harbor, New York

Sudo K, Chinen K, Nakamura Y (1994) 2058 expressed sequence tags (ESTs) from a human fetal lung cDNA library. Genomics 24:276–279

Watson CJ, Demmer J (1995) Procedures for cDNA cloning. In: Glover DM, Hames BD (eds) DNA cloning: a practical approach, vol 1, 2nd edn. IRL Press, Oxford, pp 85–119

Chenchik A, Zhu, YY, Diatchenko L, Li R, Hill J, Siebert PD (1998) Generation and use of high quality cDNA from small amounts of total RNA by SMART PCR. In: Siebert PD and Larrick J (eds) Gene cloning and analysis by RT-PCR. Biotechniques Books Natick, MA, pp 305–319

## ▥ Suppliers

BD Biosciences Clontech, 1020 East Meadow Circle, Palo Alto, California 94086–4230, USA (e-mail: custsvc@clontech.com, Tel.: +1-800-6622566, Fax: +1-800-4241350, http://www.clontech.com)

Epicentre Technologies, 1462 Emil Street, Madison, Wisconsin 53713, USA (Tel.: +1-800-2848474, Fax: +1-800-2583088, http://www.epicentre.com)

Life Technologies, 8400 Helgerman Cort, P.O. Box 6009, Gaithersburg, Marykabd 20898–9980, USA (e-mail: info@lifetech.com, Tel.: +1-800-8408000, Fax: +1-800-3312286, http://www. lifetech.com)

New England Biolabs, 32 Tozer Road, Beverly, Massachusetts 01915–9965, USA (e-mail: orders@neb.com, Tel.: +1-800-6325227, Fax: +1-508-9211350, http://www.neb.com)

Perkin Elmer, 850 Lincoln Centre Drive, Foster City, California 94404, USA (Tel.: +1-415-5706667, Fax: +1-415-5722743, http://www.perkinelmer.com)

Pharmacia Biotech Inc., 800 Centennial Ave. Piscataway,
New Jersey 08855–1327, USA
(e-mail: CatMaster@eu.pharmacia.com, Tel.: +1-800-5263593,
Fax: +1-800-3293593, http://www.biotech.pharmacia.com)

QIAGEN Inc., 28159 Avenue Stanford, Valencia,
California 91355, USA (Tel.: +1-800-4268157,
Fax: +1-800-7182056, http://www.qiagen.com)

SIGMA Chemical Company, P.O. Box 14508, St. Louis,
Missouri 63178, USA (e-mail: custserv@sial.com,
Tel.: +1-00-3253010, Fax: +1-800-3255052,
http://www.sigma. sial.com)

# Subtractive and Differential
# Display Techniques

# Subtractive Hybridization and cDNA Cloning

R. CURTIS BIRD and GIN WU

## ▦ Introduction

There has long been a desire among scientists, investigating the subtle changes that occur in cells as they grow and develop, to identify with precision, the transcripts that are newly transcribed as a result of altered physiology. Whether in response to a natural phenomenon such as a growth factor or morphogen stimulation, or in response to environmental stimulation or stress, new transcripts appear as a result of newly stimulated gene expression. Often these genes are best characterized as a reaction to the specific stimulus, but frequently they are the regulatory elements whose products participate in eliciting the cellular changes observed. Identification of genes of this nature, and their products, is of wide interest and applicability and accounts for the intense interest in techniques that allow a bias-free sampling of uniquely transcribed products. The difficulty has been in attaining a bias-free sample of unique transcripts in sufficient quantities to allow efficient recombinant library construction. Most techniques are anything but bias-free and so will result in the recovery of differing, though frequently overlapping, subsets of transcripts. Application of the appropriate technique requires knowledge of what biases each selection technique is dependent on and other weaknesses inherent in its application. We have

R. Curtis Bird (e-mail: birdric@vetmed.auburn.edu, Tel.: +1-334-8444539, Fax: +1-334-8442652)
Department of Pathobiology, Auburn University, Auburn, Alabama, USA 36849–5519, USA
G. Wu (Tel: +1-415-8830184, Fax: +1-415-8833037)
Cyberdent, Inc., 100 Galli Dr, Suite 9, Novato, California 94949, USA

Springer Lab Manual
R.C. Bird, B.F. Smith (Eds.) Genetic
Library Construction and Screening
© Springer-Verlag Berlin Heidelberg 2002

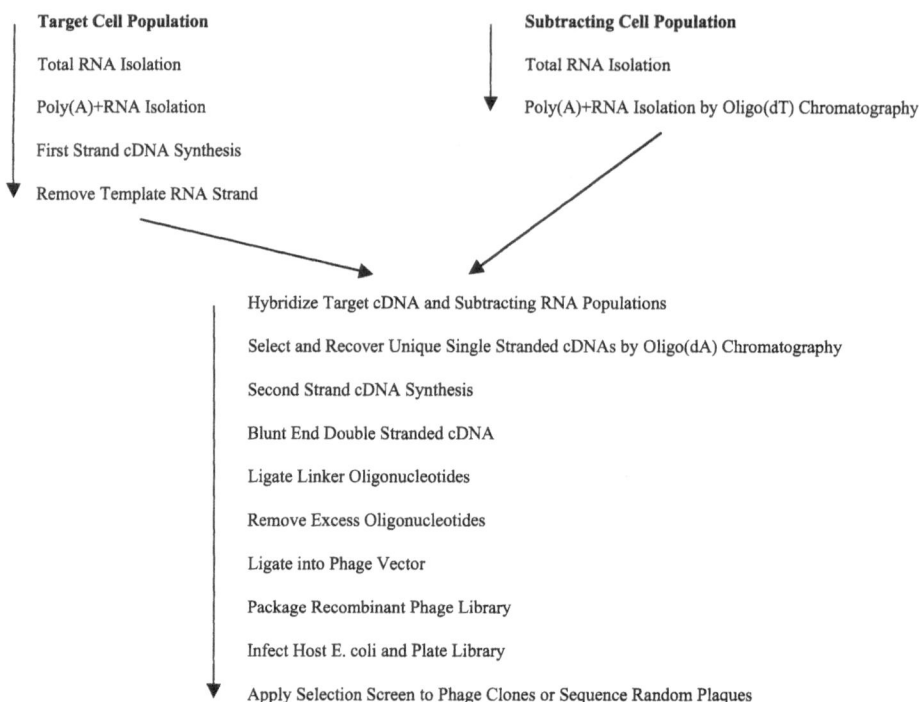

**Fig. 1.** Schematic overview of the subtractive hybridization cDNA library cloning strategy

developed techniques for the application of subtractive hybridization to synthesize a high $R_o t$ fraction cDNA library enriched in sequences expressed during physiological states or conditions unique to a particular cell. In our case this approach has been applied to cells selected for different phases of the cell cycle in human tumor-derived cell lines (Wu et al. 1993a). As is true in many other applications, the procedures for handling and selecting the cells are specific to our approach and embody biases regarding this selection. For example, the method employed for preparing synchronous populations of cells in specific cell cycle phases in sufficient quantity to prepare mRNA populations for subtraction is a critical choice. Most drug-induced synchrony regimes result in cells that are incompletely or partially synchronous. They are usually synchronous for mitosis and/or DNA synthesis but not for accumulation of cell mass which continues during the block (Mitchison 1971). This results in an abnormal cell which can divide synchronously at least once but that undergoes other changes and adjustments as a result of the induced asyn-

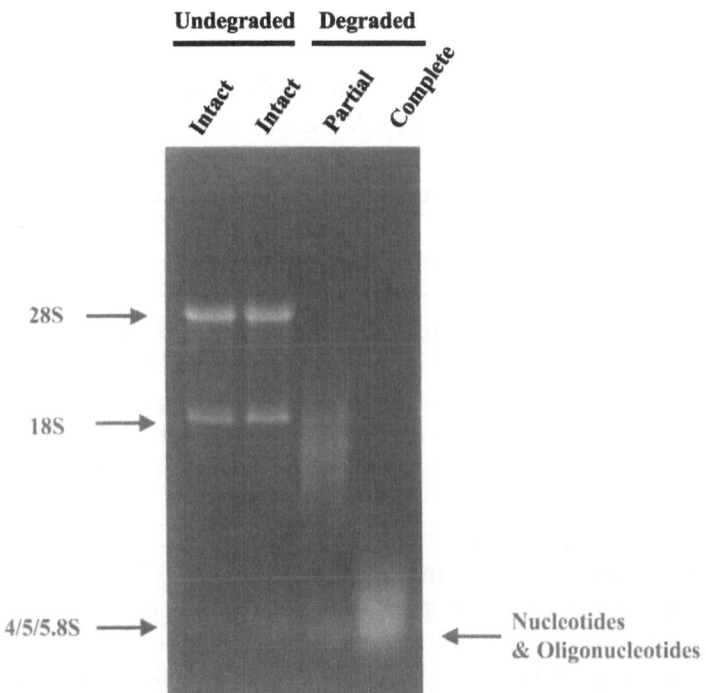

**Fig. 2.** Analysis of RNA stability under different hybridization conditions. The level of degradation present in several samples of RNA was tested using HeLa total cellular RNA and analyzed by agarose gel electrophoresis. Evidence of RNase contamination/activity can be easily detected through analysis of RNA integrity on denaturing agarose gels by comparing the staining intensity of the 28S and 18S ribosomal RNA bands (see description in the text). The positions of 28S and 18S rRNAs as well as 4S tRNA/5 or 5.8S rRNAs are indicated, as is the presence of excess oligonucleotides generated by degradation (*arrows*). The staining intensity should be approximately 2:1 and not less as the amount of ethidium bromide bound is proportional to nucleic acid length and 28S rRNA is approximately twice the length of 18S rRNA. Compare intact lanes to the partially degraded lane. The presence of excess material at the dye-front is also an indication of extensive degradation though the presence of 4S and 5/5.8S RNAs very near the front can make this determination in conditions resulting in light contamination cases less certain (compare the material detected at the front in lanes containing partial and completely degraded RNA samples). Though oligonucleotides, resulting from degradation, run approximately at the same position as tRNA, they tend to extend above and below the tRNA band

chrony between mass accumulation and DNA synthesis or mitosis. Such problems can greatly affect the expression of genes that might be anticipated during a normal continuous cell cycle resulting in a potential artifact. We chose to select our cells by centrifugal elutriation due to the quality of the synchrony, ease and speed of preparation, and lack of trauma associated with this technique (Bird 1998a,b). Once phase-specific fractions have been acquired, very high quality populations of relatively synchronous cells can be returned to culture for analysis. Total RNA can then be purified in a sufficient quantity to allow subtraction of essentially 98–99 % of common sequences, still leaving sufficient RNA mass for the cloning of the library (Fig. 1). This last aspect is of particular importance when considering that poly(A)$^+$RNA must be isolated prior to subtraction. The quality of the RNA populations is characterized by high purity of nucleic acid ($OD_{260}/OD_{280} \geq 2.0$) and highly intact and undegraded molecules (as assessed by agarose gel electrophoresis; Fig. 2). Once two high quality populations of RNA which are to be compared, have been isolated, the subtraction steps can begin.

The subtraction steps also include several critical characteristics that must be optimized for successful subtraction to be achieved. The efficiency of subtraction, integrity of residual single stranded cDNA, and efficient recovery of ng-quantities of cDNA are the three most important factors affecting quality of subtractive hybridization reactions prior to subtractive cDNA library construction. Techniques for efficient isolation of single-stranded (ss) cDNA, after subtraction, have greatly improved from early protocols based on hydroxylapatite chromatography to phenol/chloroform extraction of biotin/streptavidin cross-linked polynucleotides or oligo(dA)-cellulose affinity chromatography (Davis et al. 1984; Hazel and St. John 1988; Batra et al. 1991; Wu et al. 1993a). Continuous improvements in separation strategy have improved the purity and recovery of the ss-cDNA population of interest and allowed efficient recovery of ever smaller amounts of cDNA. These improvements have allowed the practical limits on $C_0t$ values for such reactions to be extended considerably. Factors affecting mRNA stability at the hybridization step, however, also have consequences which directly affect the complexity of the library and the length of cDNAs recovered. We have optimized the subtractive hybridization step in the cDNA cloning protocol to ensure that ss-cDNAs survive hybridization as near to full length as possible. These improvements have enabled successful construction of subtractive cDNA libraries from the ng-quantities of ss-cDNA remaining after extensive liquid hybridization to high calculated $C_o t$ values. Using this approach, a G1 phase subtractive cDNA library was made by subtracting G1 phase cDNA with a tenfold excess of S phase mRNA. Single stranded G1 phase cDNAs were then isolated by oligo(dA)-chromatography and cloned into a modified phage/bacterial plasmid vector. The library was then screened with a high $R_o t$ fraction subtractive probe population followed by screening for positive clones (Wu et al. 1993a,b).

The hybridization reaction environment has an important impact on ss-cDNA stability and clonability. Hairpin loop formation following denaturation of ss-cDNA, prior to annealing, can cause the formation of short double-stranded regions at the 5'-end which can inhibit their ligation and render such mRNAs unrecoverable resulting in their loss from the subtractive cDNA library. In addition, the length and integrity of the cDNA synthesized is dependent on the stability and integrity of the template RNA and the first strand cDNA during the hybridization reaction which often employs harsh conditions. A series of different formulations of hybridization solutions and different incubation conditions were designed and tested to optimize the subtractive hybridization protocol to allow the most efficient recovery of intact cDNA molecules possible. These reaction conditions provided optimal recovery of cDNAs from the subtractive hybridization step (Wu et al. 1993b).

It is also of great importance to ensure that the two cell populations, to be compared, be as distinct as possible and not cross-contaminated in any significant way. As little as a 10 % contamination of the subtracting cell population with the target cell population can result in essentially a complete removal of all target sequences due to the efficiency of hybridization and the abundance ratio of subtracting to target sequences (usually 10:1 or higher). Thus, if a ten-fold excess of subtracting sequence is used to select common sequences away and there is a 10 % contamination then target sequences will be present at essentially equimolar concentrations in each population. This will result in no effective selection or recovery. It is particularly a problem in cell populations where differentiation is incomplete or where separation, such as between cell cycle phases, is gradual rather than discrete. Care should be taken to minimize this problem prior to attempting subtractive hybridization. Where doubt exists, further efforts to define or analyze the cell populations to be used are necessary.

The selection regimen described below, by necessity, embodies several selection biases. The most important, however, was the assumption that some regulatory proteins would be enhanced in expression in one cell population over another. Only those cDNAs exhibiting enhanced expression levels are likely to be selected by this cloning strategy. If such molecules are less likely to be isolated through traditional strategies and differential expression is known or suspected to occur, then the subtractive hybridization approach described offers distinct advantages despite the inherent biases it imposes.

## Outline

The general scheme for synthesis and selection of the subtractive cDNA library is outlined (Fig. 1). In general, mRNA populations are selected on the basis of their poly(A)$^+$ tails by affinity chromatography and labeled subtracting (unwanted or common) and target (desired or unique) mRNA sequence populations. The subtracting population of mRNAs are then reverse transcribed to single stranded cDNA, the RNA strand is removed, and the cDNA allowed to hybridize to the target sequences to remove common mRNAs. The remaining unique mRNAs from the target cell population are then removed by affinity chromatography and the recovered target RNAs cloned into very efficient phage vectors. The descriptions of the cell lines described are clearly specific for those lines and our previous investigation of their cell cycle regulatory mechanisms. The technologies and strategies are, however, applicable to any pair of cells or tissues which are sufficiently similar and yet express significant different genes, such that the unique mRNAs may be recovered and cloned using these approaches.

## ▓ Materials

All standard laboratory reagents were of reagent grade or higher quality of manufacture and were obtained from either Fisher Scientific or Sigma Co.

- α-modified Eagle's minimal essential medium (α-MEM - **Reagents** Flow Laboratories or Gibco/BRL)
- fetal bovine serum (FBS, Sigma Co.)
- antibiotics (Sigma Co.)
- 15-cm diameter culture plates (Corning Inc.)
- Econocolumns (BioRad)
- oligo(dA)-cellulose (Sigma Co.)
- oligo(dT)-cellulose (Sigma Co.)
- DNase/RNase-free glycogen (Roche Molecular Biochemicals)
- Glass and plastic centrifuge tubes in 15-, 30- and 50-ml sizes (Corning Plastics through Fisher Scientific)
- guanidinium thiocyanate (Sigma Co.)
- oligo(dT)$_{12-18}$ (Sigma Co.)
- enzymatically modified M-MLV reverse transcriptase (Superscript, Gibco/BRL)
- [α$^{32}$P]-dCTP (Dupont/New England Nuclear)
- E. coli ligase (Gibco/BRL)
- DEPC (diethylpyrocarbonate) (Sigma Co.)
- 1 × Superscript RT-buffer (Gibco/BRL)
- Speed-Vac (Fisher Scientific)
- DNase-free bovine serum albumin (BSA, Sigma Co.)
- random hexanucleotide primers (Pharmacia)
- Klenow fragment of DNA polymerase I (Gibco/BRL)
- 10 × E. coli ligase buffer (Gibco/BRL)
- E. coli ligase (Gibco/BRL)
- T4 DNA polymerase(Gibco/BRL)
- Eco RI/Not I linkers (Invitrogen)
- 1 × T4 DNA ligase buffer (Gibco/BRL)
- T4 DNA ligase (Gibco/BRL)
- T4 polynucleotide kinase (Gibco/BRL)
- λgt11 phage cloning vector (Gibco/BRL)
- Gigapack Gold phage packaging kit (Stratagene)
- E. coli strain Y1088 (Stratagene)

**Buffers and solutions**    All buffers and solutions should be prepared using the highest quality reagents available with scrupulous attention to RNase-free technique and using only the purest RNase-free water. Sterilization by filtration or autoclaving is recommended except where noted.

### Binding Buffer – Oligo(dN) Chromatography
10 mM Tris-HCl, pH 7.6
1 mM EDTA
0.5 % SDS

### DEPC (diethylpyrocarbonate) treated and sterile water
Water is treated to be RNase-free and sterile by adding DEPC at 0.1 % v/v per liter, shaken and allowed to stand followed by autoclaving for about 30 min at 121 °C for adequate pre-treatment. Ensure no contact between skin and these items occurs prior to use, as DEPC is a suspected carcinogen.

### 10 × *E. coli* ligase buffer
400 mM Tris-HCl, pH 8.0
100 mM $MgCl_2$
50 mM DTT
0.5 mg/ml DNase-free BSA
1.5 mM β-$NAD^+$

### Elution Buffer – Oligo(dN) Chromatography
1 mM EDTA, pH 8.0
0.1 % SDS

### Ethidium Bromide Stock Staining Solution (1000 ×)
0.2 mg/ml ethidium bromide in water
Store at 4 °C in the dark (foil wrapped container). Do not sterilize. Avoid direct contact with skin. Exercise caution as this reagent is a potent mutagen and suspected carcinogen. Dispose of using appropriate procedures as toxic waste.

### Guanidinium Cell Lysis Solution
4 M guanidinium thiocyanate
25 mM sodium citrate pH 7.0
0.5 % Sarcosyl
0.1 M β-mercaptoethanol

**Prehybridization Buffer**
50 % deionized formamide
5 × SSC
0.2 % w/v polyvinylpyrrolidone
0.2 % w/v Ficoll
0.2 % w/v BSA
0.8 mg/ml yeast RNA
1 % SDS
0.1 mg/ml polyadenylic acid
10 mM Tris-HCl, pH 7.4

**5 × -Random Primer Extension Buffer**
250 mM Tris-HCl, pH 7.5
50 mM $MgCl_2$
5 mM DTT
0.25 mg/ml DNase-free bovine serum albumin (BSA)

**5 × SSC**
750 mM NaCl
75 mM sodium citrate, pH 7.0

**Subtractive Hybridization Buffer**
0.5 M HEPES, pH 7.0
2 M NaCl
5 mM EDTA
1.25 % SDS

**10 × T4 PNK Buffer**
500 mM Tris-HCl, pH 7.5
100 mM $MgCl_2$
50 mM DTT
10 mM ATP
500 µg/ml BSA

**TAE Buffer**
40 mM Tris-acetate, pH 8.5
1 mM $Na_2EDTA$

**1 × TE Buffer**
10 mM Tris-HCl, pH 8.0
1 mM $Na_2EDTA$

**1 × TES Buffer**
10 mM Tris-HCl, pH 7.4
1 mM Na₂EDTA
0.5 % SDS

**General considerations**

In general, laboratory conditions should be maintained such that the potential for contamination of the reactions by environmental or introduced RNases is kept to a minimum. Due to the length of the reactions, the high reaction temperatures and the robust nature of RNase activity, it is paramount that introduction of this activity be essentially eliminated if possible. Such precautions include limiting contact with skin or aerosols (particularly human derived) and isolating these activities to one area of the laboratory not used for other purposes. Though not essential, in cases where RNase has been a problem such isolation can help. Ultimately, scrupulous attention to RNase-free technique and thorough decontamination of glass and plastic ware are the most important steps to prevent contamination. Bake glassware at 200 °C for at least 4 h under dry conditions and treat as sterile for bench-top use. Most plastic ware can be autoclaved for about 30 min at 121 °C, after immersion in DEPC-water, for adequate pre-treatment. Ensure no contact between skin and these items occurs prior to use. The use of gloves is not essential for skilled/experienced investigators in the handling of RNA but can help those with less experience. As in all applications, gloves do not make up for a lack of appropriate RNase containment and should be removed whenever they become contaminated (e.g., touched by bare skin or the bench top), require refreshing, or when the investigator leaves the RNase-free area of the laboratory. Evidence of RNase contamination can be easily detected through analysis of RNA integrity on denaturing agarose gels by comparing the staining intensity of the 28S and 18S ribosomal RNA bands (should be 2:1 approximately and not less as the amount of ethidium bromide bound is proportional to nucleic acid length and 28S rRNA is almost exactly twice the length of 18S rRNA; Fig. 2). The presence of excess material at the dye-front is also an indication of degradation though the presence of 4 S and 5/5.8S RNAs very near the front can make this determination in light contamination cases less certain.

## ▨ Procedure

### Cell culture and RNA isolation

Any two cell populations can be compared using these procedures. However, it is important to select these populations such
that they are as similar as possible with a minimum of cross-
contamination (see above). The example offered here is taken
from our efforts to compare different phases of the cell cycle
and provide an example of both very similar populations as
well as the obvious problems of cross-contamination due to
limitations in the ability to separate these populations effectively (Bird 1998a,b). Because they are not germane to the sub-
ject of subtractive hybridization and because they have been
described in detail elsewhere, the reader is directed to these ref-
erences for a detailed discussion of separation of cells based on
cell cycle phase.

1.  HeLa S3 cells were cultured in 20 ml of modified Eagle's
    medium α-MEM with 10 % fetal bovine serum (FBS) and
    antibiotics (Sigma Co.) in 15-cm diameter culture plates
    (Corning Inc.) at 37 °C with 5 % $CO_2$ and 100 % humidity
    (Bird et al. 1988, 1990).

2.  Cells were collected at 60–70 % confluency by trypsin diges-
    tion. An appropriate growth rate and cell condition/treat-
    ment for the subtractive comparison should be employed.
    Rapid exponential growth is required for cell cycle analysis.

3.  Cells were concentrated by low speed centrifugation at 4 °C
    and resuspended in 5 ml of ice cold α-MEM with 5 % FBS for
    every three plates.

4.  Synchronous G1 phase populations of cells were obtained by
    centrifugal elutriation at a loading speed of 15 ml/min fol-
    lowed by a stepwise increase in eluting flow rate from 23 to
    43 ml/min.

5.  A portion of each of the synchronous cell fractions were fixed
    in 70 % ethanol for flow cytometric analysis and the remain-
    der of the cells were immediately prepared for RNA isolation
    (Bird 1998a,b).

### Isolation of mRNA

Total RNA was isolated by the guanidinium isothiocyanate procedure as previously described (Chomczynski and Sacchi 1987) though any method can be used that preserves the integrity of the RNAs and provides pure samples. Poly(A)⁺RNA was isolated by three rounds of binding and elution to oligo(dT)-cellulose (Favaloro et al. 1980; Bird et al. 1985). RNA stability in different hybridization buffers and at different temperatures and incubation times was analyzed by 1 % w/v agarose gel electrophoresis in TAE buffer followed by staining in 0.2 µg/ml ethidium bromide (Sambrook et al. 1989). In summary, the best conditions found for the liquid subtractive hybridization of subtracting cDNAs to target mRNAs are as follows: 42 °C in the presence of 50 % v/v formamide, 50 mM HEPES, pH 7.6, and 0.5 % w/v SDS for up to 48 h (Wu et al. 1993b).

1.  Collect cells of choice by low speed centrifugation (as appropriate) and lyse the cell pellet in guanidinium cell lysis solution (100 µl of solution per $1 \times 10^6$ cells).

2.  Add the following reagents at room temperature mixing each reagent well immediately after each addition.

    0.1 volumes of 2 M sodium acetate

    1 volume of water-saturated phenol

    0.2 volumes of chloroform

3.  Centrifuge each sample at 10,000 $g$ for 20 min at 4 °C.

4.  Transfer the aqueous phase to a clean centrifuge tube.

5.  Precipitate RNA by addition of 1 volume of isopropanol and incubation at –20 °C for at least 1 h.

6.  Collect the RNA by centrifugation at 10,000 $g$ for 20 min at 4 °C.

7.  Resuspend in 0.3 volumes of guanidinium cell lysis solution.

8.  Reprecipitate the RNA by adding an equal volume of isopropanol and incubation at –20 °C for 1 h.

9.  Centrifuge at 10,000 $g$ for 10 min at 4 °C.

10. Dissolve the RNA pellet in 0.5 ml TES buffer.

    Poly(A)⁺RNA was purified by repeated oligo(dT)-cellulose column chromatography by three rounds of binding and elution.

11. Prepare a disposable chromatography column of approximately 0.7 × 10 cm (fitted with a three-way stopcock - BioRad Econocolumn) by soaking the column in 1.0 M NaOH for 10 min. A syringe fitted to the bottom of the column can be used to force the bubbles of air out of the column support frit ensuring complete wetting of the column components. Then, rinse copiously with sterile/RNase-free water.

12. Suspend 0.2 g oligo(dA)-cellulose in excess binding buffer (~10 ml) in a 15-ml disposable centrifuge tube (Corning).

13. Wash twice by allowing the cellulose to settle and decanting the buffer followed by resuspension in fresh binding buffer. Avoid production of fines.

14. Pipette the suspended slurry into the column and allow to settle at unit gravity.

15. Run the residual buffer out of the column. The meniscus will stop the flow at the top of the column, which will not run dry unless left to desiccate.

16. Total RNA (1–2 mg) is suspended in binding buffer (without salt), relaxed by heating to 65 °C for 1 min, and then made up to 0.5 M NaCl from a 5 M stock solution, mixed thoroughly, and quenched in an ice-bath until it cools to about room temperature (no colder or the SDS will precipitate).

17. Apply the RNA/binding buffer solution to the column and then allow it to run through the cellulose bed at approximately 1 drop/min. Repeat application of the eluate two additional times. The resulting solution, absent of any RNAs that have stuck to the column is considered stripped of poly(A)⁺RNAs and is termed the poly(A)⁻ fraction.

18. Wash the column with approximately 20 ml binding buffer.

19. Elute the poly(A)⁺RNAs with 4 ml elution buffer into a 30-ml baked Corex centrifuge tube.

20. Precipitate by addition of 2 ml (0.5 volumes) 7.5 M ammonium acetate and 12 ml (3 volumes) ethanol followed by incubation at –20 °C overnight.

21. Collect the RNA pellet by centrifugation at 10,000 $g$ for 20 min at 4 °C. Drain well by inversion and wipe rim of inverted tube to remove residual ethanol/salt solution. Allow to air-dry so that the pellet appears just damp to the eye and redissolve in 0.5 ml sterile water or TE buffer.

22. The composition of the RNA in such a sample is approximately 1:1 poly(A)$^+$ to poly(A)$^-$ RNAs as a result of removing approximately 90% of the RNA in this first affinity chromatography step. Subtraction procedures usually require higher levels of purity however, and essentially pure poly(A)$^+$RNA can be recovered by repeating steps 14–19 two additional times. The resulting RNA will be essentially free of contaminating poly(A)$^-$ sequences.

23. Store at –20 °C (short term) or –85 °C (long term) until needed.

24. Determine the concentration and purity of each sample by diluting 5 µl of RNA solution in 500 µl water and determining the optical density (absorbance) at 260, 280 and 320 nm light.

25. Calculate concentration using the following formula.

RNA concentration (µg/ml)=[$OD_{260}$–$OD_{320}$] × 500/5 × 40 µg/$OD_{260}$ unit

Where concentration is the corrected optical density × dilution factor × constant

The constant is calculated to provide a concentration in µg/ml provided a standard 13 × 13 mm quartz glass cuvette with 1-mm-thick walls is used for the determination (path length of light is 11 mm through the sample).

Apparent optical density in the far visible (320 nm) is subtracted from all values collected in the ultraviolet part of the spectrum to eliminate apparent optical density due to non-wave length-specific optical density resulting from undissolved particles in suspension or obstructing finger prints or bubbles if present. $OD_{320}$ should ideally be zero in well prepared samples in very clean cuvettes.

26. Calculate purity as $[OD_{260}-OD_{320}] / [OD_{280}-OD_{320}] \geq 1.8$ (and preferably 2.0)

27. Estimate intactness of each RNA sample by agarose gel electrophoresis, as described in the introduction, in a denaturing buffer system such as MOPS/formaldehyde (Ausubel et al. 1991).

28. High quality poly(A)+RNA preparations should yield approximately 5 % (up to 100 µg from 2 mg total RNA) poly(A)+RNA of OD ratio 2.0 or higher.


## Subtractive cDNA library construction

cDNA synthesis and subtractive hybridization were performed as previously described with modification (Wu et al. 1993a,b). The G1 phase-specific subtracted cDNA library was constructed by a modification of the method originally described by Batra et al. (1991). Using poly(A)+RNA purified from centrifugal elutriated G1 phase HeLa S3 cells as the template and oligo(dT)$_{12-18}$ as the primer, the first strand G1 phase cDNA was synthesized using enzymatically modified M-MLV reverse transcriptase (Gibco/BRL-Superscript). Following hybridization, ss-cDNAs were isolated by oligo(dA)-cellulose column chromatography (Batra et al. 1991). Second-strand cDNAs were synthesized by random primer extension including [$\alpha^{32}$P]-dCTP and the nicks sealed with E. coli ligase (Gibco/BRL) (Sambrook et al. 1989). cDNA synthesis and size at each step were analyzed by alkaline agarose gel electrophoresis and autoradiography as previously described (McDonell et al. 1977).

In several attempts to construct this library, 3 µg of G1 phase poly(A)+RNA template produced approximately 15 ng of labeled G1 phase cDNA after subtraction. The full length and intact nature of both first- and second-strand cDNA reaction products was confirmed by the continued presence of full-length cDNAs following both reactions (Fig. 3). If strict control of RNase-free conditions are maintained, little evidence of degradation or reduction in size of cDNAs during either first- or second-strand synthesis was observed (Fig. 2). The resulting subtraction efficiency was about 97 % and the theoretical efficiency was calculated to be approximately 99.6 % (Hames and Higgins 1985).

1ˢᵗ Strand Synthesis          1ˢᵗ Strand Synthesis          2ⁿᵈ Strand Synthesis
                             After NaOH Hydrolysis

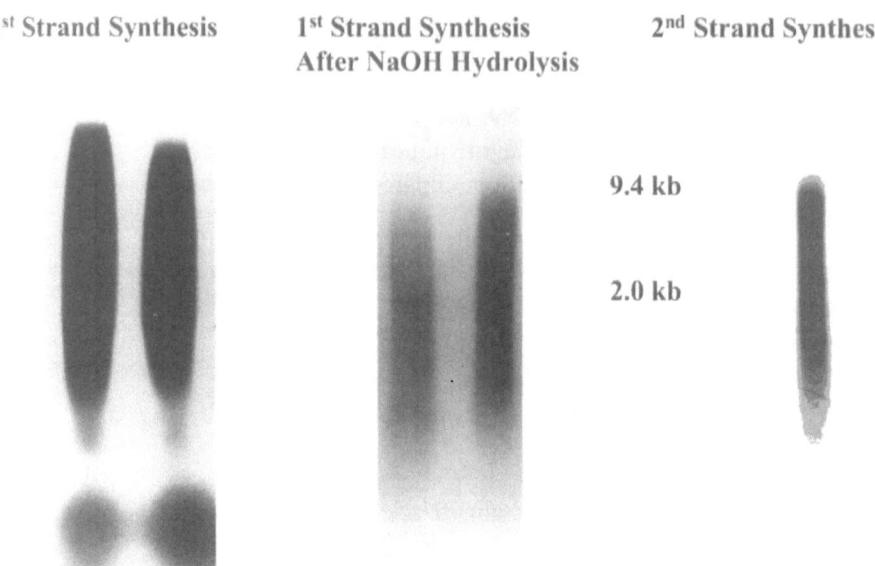

9.4 kb

2.0 kb

**Fig. 3.** Analysis of cDNA integrity following first- and second-strand cDNA synthesis reactions. cDNAs, labeled with [α³²P]-dCTP, were analyzed by alkaline agarose gel electrophoresis and autoradiography. *1st Strand Synthesis* First-strand cDNA reactions; *1st Strand Synthesis After NaOH Hydrolysis* first-strand cDNA reaction following removal of the RNA template with NaOH; *2nd Strand Synthesis* second-strand cDNA reactions. Note that the maximum and minimum sizes of the cDNA populations in each reaction are essentially the same and that the proportion of molecular weight distribution is also essentially the same at each step of the reaction. This indicates that most cDNAs survive essentially intact through each of the synthetic and hybridization steps of the cloning reactions where conditions are the harshest. The position of molecular weight markers is noted

1. First-strand cDNA synthesis reaction in a total of 50 µl.

   To a siliconized and DEPC (diethylpyrocarbonate) treated 0.5-ml microfuge tube add:

   0.5–3.0 µg of poly(A)⁺RNA from the target cell population

   1 × Superscript RT-buffer (Gibco/BRL)

   0.5 mM dNTPs

   1 mM DTT

   0.05 µg/µl oligo(dT)$_{12-18}$

   500 units M-MLV reverse transcriptase (Gibco/BRL)

2.  Immediately after the first strand reaction mix is prepared, the following components are transferred to a separate tube as a tracer reaction.

    10 µl of the reaction mix from step 1

    1 µCi [$\alpha^{32}$P]-dCTP

    Both reaction tubes were incubated at 37 °C for 1 h, chilled on ice, and the tracer reaction was analyzed by alkaline agarose gel electrophoresis (Sambrook et al. 1989).

3.  In the main 40 µl reaction tube, the template RNA was digested by alkaline hydrolysis with the addition of an equal volume of alkaline hydrolysis buffer

    1 volume of 0.6 M NaOH

    2 mM Na$_2$EDTA

    Incubate at 65 °C for 30 min.

4.  Digestion was stopped/neutralized and nucleic acids precipitated by chilling on ice followed by the immediate addition of:

    1.4 µl 2.0 M HEPES, pH 7.5

    1.4 µl sodium acetate, pH 4.0

    1 µl 20 µg/µl DNase/RNase-free glycogen (Roche Molecular Biochemicals)

    100 µl ethanol.

    cDNA was precipitated by incubation at –70 °C for 30 min before it was collected by centrifugation, at 4 °C, for 30 min (10,000–14,000 g).

5.  The cDNA was resuspended in 30 µl of DEPC-treated water containing 5–30 µg of the poly(A)$^+$RNA isolated from the subtracting RNA population (approximately ten times the mass of input target RNA).

6. The sample was dried down in a Speed-Vac and resuspended in:

   1 µl of DEPC-treated water

   4 µl of subtractive hybridization buffer

7. Subtractive hybridization was carried out by adding

   5 µl of deionized formamide

   15 µl of RNase-free mineral oil as an overlay

8. The reaction was incubated at 42 °C for 24 h (theoretical completion of subtraction was >99 %).

9. The remaining excess single-stranded cDNA in the reaction was isolated by two rounds of oligo(dA)-cellulose chromatography. This procedure is performed exactly as described for oligo(dT) chromatography described above.

10. The subtracted and selected target single-stranded cDNA population was co-precipitated by adding (1 volume is the eluate volume recovered off the oligo(dA) column):

    20 µg of glycogen in 1/2 volume 7.5 M ammonium acetate

    3 volumes ethanol

    Subtracted cDNA was precipitated by incubation at −70 °C for 30 min before it was collected by centrifugation, at 4 °C, for 30 min (10,000–14,000 $g$).

11. Double-stranded cDNA was synthesized by resuspending the ss-cDNA, precipitated after the subtraction reaction, in 20 µl DEPC-treated water and adding:

    1 µl of 10 mM dNTPs

    6 µl of 5 × random primer extension buffer

    1 µl of 1.5 µg/µl random hexanucleotide primers (Pharmacia)

    2 µl of 10 unit/µl Klenow fragment of DNA polymerase I

12. Immediately, 5 µl of the mixture was transferred to a second tube containing 1 µl (10 µCi) of $[\alpha^{32}P]$-dCTP as the tracer reaction.

13. The reaction mixtures were incubated at room temperature for 2 h and then transferred to 37 °C for 30 min. The sample was heated to 70 °C for 10 min to denature the enzyme.

14. The now double stranded cDNA was co-precipitated with glycogen.

    20 µg of glycogen in 1/2 volume 7.5 M ammonium acetate

    3 volumes ethanol

    The cDNA was precipitated by incubation at –70 °C for 30 min before it was collected by centrifugation, at 4 °C, for 30 min (10,000–14,000 $g$).

15. After centrifugation the ds-cDNA was dissolved in:

    8 µl of water

    1 µl of 10 × E. coli ligase buffer

    1 µl E. coli ligase

    The solution was mixed and incubated at 16 °C for 30 min.

16. The cDNA was phenol/chloroform extracted.

    1 volume equilibrated/buffered phenol was added and vortexed for ~15 s.

    The mixture was separated by centrifugation for 1 min (10,000–14,000 $g$).

    1 volume chloroform was added and vortexed for ~15 s.

    The lower, organic layer was removed with a micropipetter

    1 volume chloroform was added and vortexed for ~15 s.

    The mixture was separated by centrifugation for 1 min (10,000–14,000 $g$).

    The lower, organic layer was removed with a micropipetter

    1 volume chloroform was added and vortexed for ~15 s.

    The mixture was separated by centrifugation for 1 min (10,000–14,000 $g$).

    The upper, aqueous layer was carefully removed with a micropipetter to a fresh tube

17. The cDNA was precipitated by addition of ethanol and salt

    20 µg of glycogen in 1/2 volume 7.5 M ammonium acetate

    3 volumes ethanol

    The cDNA was precipitated by incubation at –20 °C for 1 h before it was collected by centrifugation, at 4 °C, for 30 min (10,000–14,000 $g$).

18. The cDNA was blunt-ended by first dissolving the pellet in (final volume of 80 µl):

    50 µl of DEPC-treated water

    0.3 mM dNTPs

    30 mM Tris-acetate, pH 7.8

    67 mM potassium acetate

    10 mM magnesium acetate

    87.5 µg/ml RNase/DNase-free BSA

    100 unit/ml T4 DNA polymerase

    The blunt-ending reaction mixture was incubated at 37 °C for 45 min followed by phenol/chloroform extraction as described above (step 15).

19. The blunt-ended cDNA was precipitated from ethanol as described above and resuspended in 1 × TE buffer (approximately 50 µl).

20. Eco RI/Not I linkers (Invitrogen) were covalently added to the cDNA termini by addition of (total volume of 10 µl):

    1 µg linkers

    1 × T4 DNA ligase buffer (Gibco/BRL)

    10 units T4 DNA ligase (Gibco/BRL)

    The reaction was incubated at 7 °C overnight.

21. Excess linkers were removed by precipitation of the cDNA by addition of:

    1/10 volume 3 M sodium acetate, pH 6.0

    0.6 volumes isopropanol

    The mixture was chilled on ice for 15 min and centrifuged for 5 min at 14,000 $g$ and 4 °C.

22. The pellet was washed gently with:

    1/2 of the original volume of 40 % isopropanol

    1/10 of original volume of 3 M sodium acetate, pH 6.0

23. The DNA was briefly dried under vacuum and resuspended in approximately 50 µl 1 × TE buffer.

24. The linker modified ends were phosphorylated with T4 polynucleotide kinase (Ausubel et al. 1991).

    50 µl linker modified double stranded cDNA in TE buffer

    10 µl 10 × T4 PNK buffer

    20 units T4 polynucleotide kinase

    DEPC-water to 100 µl total volume

    The reaction was incubated at 37 °C for 60 min and the reaction was stopped by addition of 1 µl 0.5 M EDTA. The final phosphorylated cDNA was stored frozen at –85 °C in anticipation of ligation into an appropriate phage-based cloning vector.

25. The cDNA was inserted into a λgt11 vector using T4 DNA ligase as described above for linker addition (Sambrook et al. 1989).

26. The reconstituted phage genomes incorporating the inserted cDNAs were then packaged into phage particles in vitro utilizing a commercial packaging kit (Gigapack Gold, Stratagene). E. coli strain Y1088 was employed as the host strain for plaque growth and screening.

### Screening of the cDNA library with subtracted probe populations

The G1 phase subtracted probe population to be used to screen for specific cDNA plaques, was synthesized by incorporation of [$\alpha^{32}$P]-dCTP into G1 phase cDNA population followed by subtraction with S phase poly(A)$^+$RNA by slightly modifying the protocol described above.

1.  The high specific-activity first strand G1 phase cDNA was synthesized substituting:

    0.5 µg target (G1 phase) poly(A)$^+$RNA template

    1 mCi of 800 Ci/mmol [$\alpha^{32}$P]-dCTP as labeling substrate.

2.  The labeled target G1 phase ss-cDNA was then subtracted with 5 µg of S phase poly(A)$^+$RNA for 3 days under the conditions described for library construction (see above).

3.  The labeled and subtracted target G1 phase-specific ss-cDNA was isolated by oligo(dA)-cellulose chromatography as described for poly(A)$^+$RNA isolation (see above).

4.  The subtracted G1 phase cDNA library was plated on a total of 16 15-cm diameter 2YT plates, with each of the plates containing about $5 \times 10^5$ plaques (Sambrook et al. 1989).

5.  Plaque replicas of each plate were lifted on nylon membranes and lysed, fixed and denatured as described (Sambrook et al. 1989).

6.  Each of the membranes was prehybridized with 50 ml of prehybridization buffer at 42 °C for 3 h.

7.  The subtracted probe was then added to the prehybridization solution at $4 \times 10^6$ cpm/10 ml/filter, and the hybridization reaction was incubated at 42 °C for 48 h with rotation.

8.  Membranes were washed twice with excess $2 \times$ SSC, 1 % SDS at room temperature for 20 min followed by two washes in $0.1 \times$ SSC, 0.1 % SDS at 68 °C for 60 min.

9.  Membranes were dried at 37 °C and prepared for autoradiography on X-ray film.

Library screening was performed in two separate steps. Approximately $5 \times 10^5$ plaques were plated on each of 16 15-cm diameter plates and hybridized to the subtracted probe (Sambrook et al. 1989; Ausubel et al. 1991). Approximately 200 positive clones were selected and picked. The 200 plaques were mixed and replated on eight plates at a density of about 500 plaques/plate. After probing with a second G1 phase subtracted probe, 20 positive clones were selected and purified. cDNA inserts were released from the vector by restriction endonuclease digestion to release the cDNA insert followed by analysis using agarose gel electrophoresis (Wu et al. 1993a). The size of the largest insert isolated was 1.3 kb and the smallest insert was about 300 bp. The sizes of most inserts were between 600 to 700 bp. Each of the cDNA clones was released from the phage vector by restriction enzyme digestion and subcloned into a bacterial plasmid vector, pTZ19U (US Biochemicals) or equivalent. Alternatively, primers standard for the multiple cloning site of the phage cloning vector were used to PCR-amplify the insert out of the phage clones and allow subcloning by the T/A method (see Chap. 2, this Vol.) into pCR2.1 from Invitrogen (Ausubel et al. 1991). DNA sequencing of each clone recovered was performed by double-stranded DNA sequencing as previously described (Sambrook et al. 1989) or by automated DNA sequencing in the institutional core sequencing facility prior to comparison to sequences in the GenBank database using FASTA analysis (Ausubel et al. 1991). It is important to note that each clone was sequenced several times in both directions and that when PCR was used to amplify clones, multiple clones were sequenced to detect Taq polymerase-induced mutations.

**Comparison of subtracting and target mRNA expression profiles**

As stated previously, a knowledge of the degree of overlap in the two cell populations to be subtracted is critical if efficient subtraction is to be obtained without loss of the target mRNA population. Thus, poly(A)$^+$mRNA populations isolated from both the subtracting and target cell populations should be analyzed for differences and overlap/similarities in gene expression profiles. If access to an arraying system is available, this would provide a very detailed analysis of such events allowing the analysis of expression of genes known to be differentially expressed in one or the other of the two cell populations. Lack of a distinct expression signal in either population but especially contamination of the subtracting population with the target population should be cause for concern. DNA probes that are expressed equally in both populations, including α-actin or ribosomal protein L37 should also be included (Su and Bird 1995). Other means of comparing cell populations can also be employed. Flow cytometric analysis, based on measurements of DNA content and cell volume, or fractionation/discrimination based on antibody binding can be used to effectively analyze cell populations. Analysis of metabolic differences can also be informative through the use of radioisotopes such as $^3$H-thymidine incorporation (Wu et al. 1993a). Each means of analysis will offer a different window on the character of each cell population so that several means of analysis should be used in order to develop a broad picture of each cell population to be compared.

### Identification and confirmation of subtraction selected cDNA clones

To confirm the target population-specific expression of selected cDNAs, mRNA samples can be analyzed by Northern blot and hybridized with the selected cDNA and a control transcript clone such as ribosomal protein L37 or α-actin to confirm equal loading across lanes of the gel (see above). Analysis of mRNA fractions by autoradiography and densitometry will indicate whether expression is enhanced in comparison to control transcripts. Alternatively, rt-PCR can be used to analyze the expression of selected transcripts in mRNAs from the target and subtracting cell populations. Care should be taken, however, not to over-interpret the results from rt-PCR assays, particularly those utilizing more than 25 cycles of amplification, as the results are distinctly nonlinear. Where quantifiable rt-PCR assays are required, a real time analysis of PCR kinetics of amplification are necessary.

The cloning strategy recommended here does impose some additional characteristic biases on the resulting libraries and cDNA clones selected from them. For example, since the first strand cDNA synthesis reaction was primed with oligo(dT)$_{12-18}$ and the second strand cDNA synthesis reaction was primed with random oligomeric primer DNA, this imposes a bias toward the 3'-ends of the sequences recovered. It is unlikely that cDNA clones selected from these libraries will represent full length cDNA or represent many 5'-proximal sequences. However, Northern blot analysis can be used to determine full length message size for reference and then alternative strategies can be employed to recover the 5'-termini of the cDNAs if they are required.

## Troubleshooting

### Assessment of RNA stability during hybridization

Total RNA can be used to assess stability under a variety of hybridization conditions (Fig. 2). Stability is assessed by observation of RNA samples on denaturing agarose gels to determine the amount of degradation of high molecular weight RNA. In this assay, the 28 S:18S rRNA ratio should be ~1:1 in intact samples (that is assessed as 2:1 in brightness as the 28 S rRNA is approximately twice as long as 18S rRNA and thus binds approximately twice as much of the dye). This ratio will drop to less than 2:1 and even approach 1:1 in degraded samples accompanied by an increase in the presence of excess oligonucleotides running coincident with the tRNA band at the front in ethidium bromide-stained agarose gels. We have found that when hybridization conditions included incubation at 68 °C, RNAs were all severely degraded (Wu et al. 1993b). These results indicated that RNA was subjected to strong thermal degradation at 68 °C and that at this temperature a low cloning efficiency could be expected.

High buffer capacity in the hybridization reaction is recommended as it contributed significantly to protection of RNA integrity. For example, at least 50 mM HEPES, pH 7.6, is recommended, though in our hands buffer concentrations greater than 50 mM did not further improve stability. The presence of buffer also decreased the impact of formamide breakdown and acidification of the hybridization buffer during incubation. Experiments indicated that at a hybridization temperature of 42 °C with 50 mM HEPES, pH 7.6, protection was sufficient to prevent RNA degradation for up to 48 h.

Steps should be taken to minimize evaporation or condensation within reaction tubes during hybridization as this greatly affects hybridization kinetics through an alteration in concentration of reacting species and salts/formamide. An overlay of mineral oil to prevent the sample from drying out is considered desirable. Alternatively, a thermocycler with a heated lid could also be employed to equal effect and provide excellent temperature control. We have found that addition of dextran sulfate to enhance the effective concentration of hybridizing polynucleotides caused detectable degradation. Unless it is required to drive hybridization kinetics, dextran sulfate is not beneficial for the construction of cDNA libraries (Wu et al. 1993b).

## Conclusions

Subtractive cDNA cloning is a very powerful tool for the investigation and cloning of unknown genes, especially when the target message is in low abundance and differentially expressed between two cell populations. In theory, subtraction is simple, but the experimental protocol must be carefully optimized and followed rigorously. Subtractive hybridization reactions require that the quantity of mRNA used be at least 10–20-fold greater than the target cDNA. However, the required quantity of mRNA used for subtraction should be carefully titrated because a complete subtraction requires nearly a true first order reaction (Sambrook et al. 1989). As noted above, if the RNA used for subtraction is more than 5–10 % contaminated with the RNA population used as template for the synthesis of the target cDNA, mRNAs of interest will be lost during subtraction by hybridizing to the contaminating sequences in the excess RNA driver (Wu et al. 1993b).

The integrity of the RNA used for subtraction, obviously, is also very important for the effectiveness of subtraction. Though nucleic acids can withstand relatively high temperatures, however, extended hybridization periods coupled with high temperatures during subtraction can damage polynucleotides. Minor amounts of degradation do not affect hybridization of polynucleotides to blots, but can severely decrease the quality of a subtracted cDNA library if longer to full-length cDNAs are required. The conditions described here have consistently produced high quality cDNA populations for the purposes of both library construction and hybridization subtracted probe population synthesis for library screening.

*Acknowledgements.* The authors thank Ms. Patricia DeInnocentes for valuable consultations, critical reading of the manuscript and technical assistance during the course of these investigations. This work was supported by the Food Animal and Disease Research Program at Auburn University, College of Veterinary Medicine.

## References

Ausubel FM, Brent R, Kingston RE, Moore DD, Seidman JG, Smith JA, Struhl K (1991) Current protocols in molecular biology. Greene/Wiley-Interscience, New York

Batra SK, Metzgar RS, Hollingsworth MA (1991) A simple, effective method for the construction of subtracted cDNA libraries. Gene Anal Tech 8:129–133

Bird RC (1998a) Cell separation by centrifugal elutriation. In: Celis JE (ed) Cell biology: a laboratory manual, vol 1, 2nd edn. Academic Press, San Diego, pp 205–208

Bird RC (1998b) Synchronous populations of cells in specific phases of the cell cycle prepared by centrifugal elutriation. In: Celis JE (ed) Cell biology: a laboratory manual, vol 1, 2nd edn. Academic Press, San Diego, pp 209–217

Bird RC, Jacobs FA, Stein G, Stein J, Sells BH (1985) A unique subspecies of histone H4 mRNA from rat myoblasts contains poly(A). Proc Natl Acad Sci USA 82:6760–6764

Bird RC, Bartol FF, Daron H, Stringfellow DA, Riddell MG (1988) Mitogenic activity in ovine uterine fluids: characterization of a growth factor which specifically stimulates myoblast proliferation. Biochem Biophys Res Comm 156:108–115

Bird RC, Kung T-YT, Wu G, Young-White RR (1990) Variations in *c-fos* mRNA expression during serum induction and the synchronous cell cycle. Biochem Cell Biol 68:858–862

Chomczynski P, Sacchi N (1987) Single-step method of RNA isolation by acid guanidinium thiocyanate. Anal Biochem 162:156–159

Davis MM, Cohen DI, Nielson EA, Steinmetz M, Paul WE, Hood L (1984) Cell-type-specific cDNA probes and the murine I region: the localization and orientation of Ad alpha. Proc Natl Acad Sci USA 81:2194–2198

Favaloro J, Treisman R, Kamen R (1980) Transcription maps of polyoma virus-specific RNA: analysis by two dimensional nuclease S1 gel mapping. In: Grossman L, Moldave K (eds) Methods in enzymology, vol 65. Academic Press, New York, pp 718–749

Hames BD, Higgins SJ (1985) Nucleic acid hybridization. A practical approach. IRL Press, Washington, DC

Sive HL, St John T (1988) A simple subtractive hybridization employing photoactivatable biotin and phenol extraction. Nucleic Acids Res 16:10937

McDonell MW, Simon MN, Studier FW (1977) Analysis of restriction fragments of molecular weights by electrophoresis in neutral and alkaline gels. J Mol Biol 110:119–146

Mitchison JM (1971) The biology of the cell cycle. Cambridge Univ Press, Cambridge

Sambrook J, Fritsch EF, Maniatis T (1989) Molecular cloning: a laboratory manual. Cold Spring Harbor Press, Cold Spring Harbor, New York

Su S, Bird RC (1995) Cell cycle, differentiation and tissue-independent expression of ribosomal protein L37. Eur J Biochem 232:789–797

Wu, G, Su S, Kung T-YT, Bird RC (1993a) Molecular cloning of G1 phase mRNAs from a subtractive G1 phase cDNA library. Biochem Cell Biol 71:372–380

Wu G, Su S, Bird RC (1993b) Optimization of subtractive hybridization in construction of subtractive cDNA libraries. Gene Anal Tech Appl 11(2):29–33

## ◾ Suppliers

Amersham Pharmacia Biotech, Inc., 800 Centennial Avenue, Piscataway, New Jersey 08855–1327, USA (Tel.: +1-800-5263593, www.apbiotech.com)

Bio-Rad Laboratories, 2000 Alfred Nobel Drive, Hercules, California 94547, USA (Tel: +1-800-4246723, Fax: +1-510-7411000, www.discover.bio-rad.com)

Corning Inc., P.O. Box 5000, Corning, New York 14830, USA (Tel.: +1-607-9744640)

Fisher Scientific, 2000 Park Lane Drive, Pittsburgh, Pennsylvania 15275–9943, USA (Tel.: +1-800-7667000, www.fishersci.com)

Gibco/BRL, Grand Island, New York USA (Tel.: +1-800-8286686, Fax: +1-301-4318585, www.lifetech.com)

Invitrogen, 1600 Faraday Avenue, Carlsbad, California 92008, USA (Tel.: +1-800-9556288, Fax: +1-706-6037201, www.invitrogen.com)

New England Nuclear NEN Life Science Products, Inc., 549 Albany Street, Boston, Massachusetts 02118, USA (Tel.: +1-800-5512121, Fax: +1-617-4829595, www.nen.com)

Roche Molecular Biochemicals, 9115 Hague Road, P.O. Box 50414, Indianapolis, Indiana 46250–0414, USA (Tel.: +1-800-4285433, www.ibuyRMB.com)

Sigma Co., P.O. Box 14508, St. Louis, Missouri 63178, USA
(Tel.: +1-800-3255832, www.sigma-aldrich.com)

Stratagene, 11011 North Torrey Pines Road, La Jolla,
California 92037, USA (Tel.: +1-800-4245444,
Fax: +1-619-5355400, www.stratagene.com)

US Biochemicals Corporation, 26111 Miles Road, Cleveland,
Ohio 44128, USA (Tel.: +1-800-3219322, www.usbweb.com)

# Differential Display to Identify Steroid-Induced Genes in Endocrine Cells

ROBERT J. KEMPPAINEN and ELLEN N. BEHREND

## ▨ Introduction

Differential display is one of several techniques designed to identify differentially expressed or induced genes. Although it is difficult to rank the likelihood of success comparing different methods, the steadily increasing number of citations using differential display for gene identification supports the notion that the method is effective. Even with these successes, differential display is not without pitfalls and presents challenges at several steps in the procedure (Bertioli et al. 1995; Debouck 1995).

We used differential display to identify a gene in a pituitary tumor cell line (AtT-20) that is induced when cells are treated with a glucocorticoid hormone, dexamethasone (Kemppainen and Behrend 1998). Our goal is to understand the mechanism by which glucocorticoids inhibit secretion of adrenocorticotropin from the pituitary. Previous data indicate that this negative feedback effect of the steroid hormones requires transcription and translation. Therefore, identification of genes activated in pituitary cells in response to short-term exposure to dexamethasone

R.J. Kemppainen (e-mail: kempprj@vetmed.auburn.edu, Tel.: +1-334-8444425, Fax: +1-334-8445388)
Department of Anatomy, Physiology and Pharmacology, 213 Greene Hall, College of Veterinary Medicine, Auburn University, Auburn, Alabama 36849, USA
E.N. Behrend (e-mail:behreen@auburn.edu, Tel.: +1-334-8444690 Fax: +1-334-8446034)
Department of Clinical Sciences, Auburn University, Auburn, Alabama 36849, USA

Springer Lab Manual
R.C. Bird, B.F. Smith (Eds.) Genetic
Library Construction and Screening
© Springer-Verlag Berlin Heidelberg 2002

could provide a means to understanding the mechanism for feedback.

This paper describes in practical terms our use of differential display. The reader is referred to published information regarding the theoretical aspects of the technique (Liang and Pardee 1992; Liang et al. 1995) and an excellent overall guide to the method edited by its originators (Liang and Pardee 1997).

## ■ Outline

1. Treat cells with dexamethasone (Dex) or vehicle (control)

2. Collect RNA

3. Treat RNA with DNase

4. Reverse transcribe RNA using different oligo-dT primers (-C, -G, -A)

5. Perform PCR using cDNA

6. Separate radiolabeled products on acrylamide gel

7. Perform autoradiography

8. Identify candidate cDNA bands

9. Repeat experiment at least twice using new RNA to verify results

10. Excise band(s) from gel

11. Use PCR to amplify bands

12. Excise band, use it in Northern blot to confirm identification of induced RNA

13. Clone the cDNA

14. Screen clones to identify unique species

15. Use representative cDNAs in Northern blots to identify induced genes

16. Sequence cDNA, compare to data bank

17. Use cDNA as probe to obtain full-length gene from cDNA library

- Cell culture incubator
- Centrifuge (Eppendorf Model 5415C, Brinkmann Instruments)
- Sequencing apparatus and power supply (Model S2 and Model 4001P, Life Technologies)
- PCR thermocycler (GeneAmp PCR 2400, Perkin-Elmer Corporation)
- Gel dryer (Model 583, Bio-Rad Laboratories)

Equipment

All solutions and chemicals should be molecular biology grade or equivalent.

Solutions

- DMEM-FBS: Dulbecco's Modified Eagle Medium, high glucose, with L-glutamine (DMEM) containing 10 mM HEPES, 1 × antibiotic-antimycotic, and 10 % fetal bovine serum (all from Life Technologies).
- DMEM-BSA: DMEM containing10 mM HEPES, 1 × antibiotic-antimycotic, and 0.25 % (w/v) bovine serum albumin (Sigma Chemical), filter sterilized.
- Dexamethasone (Dex; Sigma), diluted to 0.1 mg/ml in 100 % ethanol and stored at –20 °C, concentration confirmed by spectrophotometry (molar extinction of dexamethasone at 239 nm=14,900, a 25 μg/ml solution has an $A_{239}$ of 0.949)
- AtT-20 cells (gift of Dr. Steven Sabol, National Institutes of Health)

Cell culture

- Phosphate-buffered saline (Life Technologies)
- TRIzol Reagent (Life Technologies)
- Tubes (1.5- and 2.0-ml, Eppendorf Biopur, Brinkman Instruments)

RNA collection

MessageClean Kit (GenHunter Corporation) contains 10 × reaction buffer, DNase I (RNase-free), 3 M sodium acetate, and diethyl pyrocarbonate (DEPC)-treated water

DNase treatment

- RNAimage Kit (GenHunter), containing 5 × reverse transcription buffer, MMLV reverse transcriptase, dNTP mixes (25 and 250 μM), one base anchored oligo dT primers (H-$T_{11}$G, -A, and -C), 10 × PCR buffer, 8 arbitrary (13-mer) upstream primers (one-base anchored primers and arbitrary

Differential display

primers all have a 5' HindIII site), control RNA, glycogen, water, and loading dye.

- α-[$^{33}$P]dATP (10 mCi/ml, >2500 Ci/mmol, Amersham Life Sciences)
- AmpliTaq DNA polymerase (Perkin-Elmer)
- PCR tubes (MicroAmp, 0.2-ml thin-walled tubes, Perkin-Elmer)

Poly-acrylamide gel electrophoresis, autoradiography, reamplification

- Gel-Mix 6 (Life Technologies)
- Tris-Boric acid-EDTA Buffer (10 × TBE, Life Technologies)
- Whatman 3MM CHR Chromatography paper (Whatman Incorporated)
- Bio-Max MR film (Eastman Kodak)
- Radtape (Diversified Biotech)
- QIAquick Gel Extraction Kit (Qiagen Inc)
- PolyATtract mRNA Isolation Kit (Promega Corp)

## Procedure

### Cell culture and treatment with Dex

1. AtT-20 cells are maintained in 75-cm$^2$ flasks in DMEM-FBS and subcultured every 5–7 days.

2. Cells used for experiments are at 60–80 % confluency.

3. For an experiment, DMEM-FBS is removed from two flasks of cells, and replaced with warm DMEM-BSA. Incubation continues for 1 h. Then 100 nM Dex (1:2548 dilution from 0.1 mg/ml Dex stock) or an equivalent volume of 100 % ethanol (vehicle) diluted in DMEM-BSA is added to a flask (replacing medium).

4. Cells in flasks are incubated for 2 h or more in a cell culture incubator. Collect RNA.

### Collection of total RNA

1. Remove flasks from incubator and place on ice. Remove medium and wash flasks once with 15 ml of cold phosphate-

buffered saline. Do not leave phosphate-buffered saline in flask for more than 5 min (cells will detach).

2. Remove phosphate-buffered saline and immediately add 4 ml TRIzol Reagent to each flask. Agitate gently (rocking platform or by hand) for 5 min at room temperature.

3. Transfer TRIzol Reagent/cell extract to several sterile 2-ml tubes, adding approximately 1.5 ml to each tube. Add 0.3 ml of mixture (v/v, 49:1) chloroform:isoamyl alcohol to each tube. Shake and incubate 3 min at room temperature.

4. Centrifuge at 16,000 $g$, 4 °C for 15 min. Carefully transfer top aqueous layer to a new 2-ml tube. Add 0.75 ml isopropanol to each tube, mix, and incubate 10 min at room temperature.

5. Centrifuge at 16,000 $g$ for 10 min at 4 °C. Discard supernatant and add 1 ml cold 75 % ethanol to pellet. Vortex and centrifuge at 7500 $g$ for 5 min at 4 °C.

6. Discard supernatant and invert tubes to allow RNA pellets to dry. Do not dry for extended periods (>20 min) as RNA will become difficult to solubilize.

7. Solubilize RNA in RNase-free water, estimate concentration using $A_{260}$.

## DNase treatment of RNA

1. To 50 µg (can use 10–50 µg) RNA in 50 µl RNase-free water, add 5.7 µl 10 × reaction buffer, and 1 µl (10 units) DNase I (use 1.5-ml tube).

2. Mix and incubate 30 min at 37 °C.

3. Add 40 µl phenol:chloroform 5:1 (v/v), vortex 30 s, incubate 10 min at 4 °C.

4. Centrifuge at 16,000 $g$, 5 min, 4 °C, carefully collect the upper aqueous layer and transfer to a new 1.5-ml tube.

5. Add to this upper layer 5 µl of 3 M sodium acetate and 200 µl 100 % ethanol. Incubate for minimum of 1 h at –80 °C.

6. Centrifuge at 16,000 g, 10 min, 4 °C. Discard supernatant, add 0.5 ml 70 % ethanol (made using RNase-free water), centrifuge for 5 min at 12,000 g, 4 °C.

7. Remove the ethanol (do not dislodge pellet). Air dry and solubilize the RNA in 10–20 μl RNase-free water. Measure the RNA concentration using $A_{260}$, aliquot RNA into multiple tubes each containing 1 μg RNA, store at –80 °C. Avoid multiple thaw/freeze cycles.

### RNA integrity

One μg of DNase-treated RNA from each treatment should be evaluated for integrity by denaturing agarose electrophoresis (Ausubel et al. 1995).

### Differential display

**Reverse transcription**

1. Thaw DNase-treated RNA obtained from each treatment (Dex and vehicle) and dilute the RNA to 0.1 μg/ml in RNase-free water.

2. Transfer 2 μl (0.2 μg) of each of the above to three 0.2-ml PCR tubes, add 9.4 μl RNase-free water, 4 μl 5 × reverse transcription buffer, and 1.6 μl 250 μM dNTP mix to each tube.

3. To each of the three tubes/treatment add 2 μl of a different one-base anchored oligo dT primer (H-$T_{11}$G, -A, or -C). For the situation where two treatments are evaluated (Dex and vehicle), you will have a total of six tubes, three from each treatment.

4. Place tubes in the thermocycler, heat at 65 °C for 5 min, then at 37 °C for 10 min. Immediately add 1 μl MMLV reverse transcriptase to each tube. Mix contents gently and replace tube in thermocycler. Continue incubation at 37 °C for 50 min, then heat to 75 °C for 5 min, then 4 °C to finish. Store tubes at 4 °C if proceeding immediately to PCR, otherwise freeze at –20 °C.

5. Do not reuse diluted RNA.

Initially, perform PCR using a single one-base anchored oligo-dT **PCR** primer with a series of different arbitrary upstream primers. Then perform PCR using the other one-base anchored oligo-dT primers in conjunction with the upstream primers. Additional sets of upstream primers can be obtained (e.g., GenHunter Co.) or designed and evaluated depending on the goals of the experiment. For our work, 80 upstream primers were evaluated with three one-base anchored primers. This resulted in 240 primer combinations and 480 samples (Dex and vehicle treated) for polyacrylamide gel analysis. From this, one induced cDNA was detected.

To evaluate a single one-base anchored oligo dT primer with eight upstream primers (eight primers are supplied in each RNAimage kit):

1.  Make master mix, as follows (volumes based on one reaction tube):

    9.95 μl water

    2 μl      10 × PCR buffer

    1.6 μl   dNTP mix (25 μM)

    0.25 μl  α-[$^{33}$P]dATP

    0.2 μl   AmpliTaq DNA polymerase

    2 μl one-base anchored oligo dT primer (2 μM, e.g., H-T$_{11}$G)

    Prepare sufficient volume of master mix for required tubes and two extra.

2.  Dispense 16 μl of master mix per 0.2-ml PCR tube.

3.  Add 2 μl of reverse transcription mix, from step 4 above. Be sure that the one-base oligo-dT primer used for the reverse transcription is the same as that added to the PCR master mix.

4.  Add 2 μl of arbitrary primer (2 μM).

5.  To evaluate RNA from AtT-20 cells treated with Dex or vehicle, reverse transcribed with H-T$_{11}$G, and then subjected to PCR using eight arbitrary primers requires 2 × 8 or 16 tubes. To eventually evaluate all three one-base anchored primers with 80 arbitrary primers requires 480 tubes.

6.  No oil is added (when using Model 2400 thermocycler).

7.  Mix tubes and centrifuge briefly.

8.  Perform PCR using 40 cycles with each cycle as 94 °C, 15 s; 40 °C, 2 min; and 72 °C, 30 s. After cycling, incubate at 72 °C for 5 min, followed by a hold at 4 °C.

9.  Add 11 µl of loading dye to each sample, mix. Store at −20 °C until electrophoresis.

### Polyacrylamide gel electrophoresis and autoradiography

1.  Pour a 6 % denaturing, polyacrylamide gel using Gel-Mix 6 (Life Technologies) in 1 × TBE buffer, allowing at least 2 h for polymerization (preferably overnight). Use a 32-tooth, 0.4 mm, 10-µl well capacity comb.

2.  When ready to load samples, first remove comb, and rinse wells with 1 × TBE buffer, then pre-run gel (60 W constant power) for 30–60 min. Rinse wells again.

3.  Heat samples for 2 min at 80 °C and load 5.5 µl sample/lane. If possible, avoid loading the outside lanes (distortion most common here).

4.  Run gel (60 W constant power) until xylene cyanole band (upper dye band) reaches the bottom (about 2.5 h).

5.  Allow gel to cool (approximately 15 min) then apply 100 % ethanol to the short plate (enhance cooling) and gently separate the plates.

6.  Lay Whatman 3MM paper over exposed gel and press paper gently onto gel.

7.  Peel gel off plate by lifting paper slowly from glass. Place gel labels (Radtape) on exposed paper surface in several locations to orient gel for autoradiography.

8.  Do not fix gel in methanol-acetic acid.

9.  Cover gel with plastic wrap (e.g., Glad Cling Wrap or Saran Wrap) and dry at 80 °C for 2 h under vacuum.

10. Remove plastic wrap and place dried gel in film cassette. Load cassette with Kodak Bio-Max MR film being certain

that emulsion side of film is directly touching the dried gel surface. No enhancing screens are used.

11. Expose film at −85 °C for minimum of 24 h (longer exposures as $^{33}$P ages).

## Selection of bands and reamplification

1. Compare banding patterns in adjacent lanes that were generated using different treatments (Dex and vehicle) and amplified using identical primer pairs.

2. Select possible differentially expressed bands for further study. Depending on experimental goal, bands that appeared increased or decreased in intensity may be chosen. Generally, larger size cDNAs, in the upper two thirds of the gel, are the best candidates for further study.

3. Repeat the experiment twice more with new RNA; perform PCR using the same primer pairs. Select bands whose differential expression are reliably reproduced.

4. Once a band(s) is chosen for further study, cut a "window" encompassing the band from the film. Align the film with the gel/paper using the Radtape images for positioning and tape the two together. Use the window as a guide to cut the gel/paper using a scalpel.

5. Place the gel/paper containing the cDNA in a 1.5-ml tube and add 100 µl water. Soak for 10 min.

6. Heat the tube at 99 °C (boiling; pierce a small hole in the top of the tube) for 15 min.

7. Centrifuge at 10,000 g for 2 min and transfer the supernatant to fresh tube.

8. Add to the supernatant: 10 µl of 3 M sodium acetate, 5 µl of 10 mg/ml glycogen (GenHunter), and 450 µl of 100 % ethanol. Incubate at −85 °C for at least 30 min.

9. Centrifuge at 16,000 g, 4 °C, 10 min. Discard supernatant, rinse pellet with 200 µl ice-cold 85 % ethanol. Centrifuge 16,000 g, 4 °C, for 5 min. Discard supernatant.

10. Allow pellet to dry and then dissolve it in 10 μl water.

11. Perform PCR using 20.4 μl of water, 4 μl of 10 × PCR buffer, 3.2 μl of dNTP (250 μM), 4 μl of specific (i.e., primer used to initially generate band) one-base anchored primer (2 μM), 4 μl of specific arbitrary primer (2 μM), 4 μl of cDNA from step 10, and 0.4 μl Amplitaq DNA polymerase. PCR conditions are the same as for PCR during differential display.

12. Analyze the products using a 1.5 % agarose gel to determine if a band of the expected size is present. Excise the band and extract the cDNA using QIAquick gel extraction kit (Qiagen).

13. Use Northern blotting (Ausubel et al. 1995) to confirm that the cDNA fragment identifies an induced RNA. AtT-20 cells are treated as above with vehicle and Dex. Total RNA is collected and poly(A)$^+$ RNA (PolyATtract mRNA isolation kit; Promega) (2–5 μg/lane) obtained to provide strong signals on the blots. Label the cDNA with $^{32}$P using the random prime method, perform the Northern analysis using standard hybridization conditions (42 °C), washing the membrane after an overnight incubation twice for 15 min each with 1 × SSC containing 0.1 % SDS at room temperature followed by one wash at 60 °C for 20 min with 0.25 % SSC containing 0.1 % SDS.

14. Wrap membranes in plastic film and expose to Kodak Bio-Max MR film using intensifying screens. Exposure time varies from 8 h to 4 days, depending on the signal strength.

15. If the cDNA identifies an induced RNA, clone the cDNA into a vector. For our studies, the PCR-TRAP cloning system (GenHunter) was used, but any PCR-based cloning system is suitable.

16. It is likely that the cDNA generated at this point consists of more than one species. Cloned products will likely be heterogeneous and require screening. Screen several (approximately ten) randomly selected clones by partial sequencing. (Other categorization methods are available, see Liang and Pardee 1997.) Clones can then be categorized and representatives chosen for further study.

17. A representative of each clone type is screened by Northern blot, again using poly(A)⁺ RNA from Dex- and vehicle-treated cells (same protocol as above).

18. Select the clone(s) that identifies induced RNA.

19. Sequence the entire clone.

20. Compare sequence to the GenBank database. The partial sequence obtained is from the 3' end of the induced gene.

21. Use the cDNA as a probe to identify the full-length gene using a cDNA library (Ausubel et al. 1995).


## Results

In our studies to identify genes induced by short-term Dex expo-     **Example**
sure in AtT-20 cells, differential display was performed using 240 different primer combinations (three different one-base an-chored primers with 80 arbitrary primers). From this, approxi-mately 20 candidate cDNAs were identified that were repro-ducibly induced in three different experiments. Each of these bands was reamplified by PCR and 19 of 20 gave a band on agarose electrophoresis. However, of these 19, only one cDNA (the differ-ential display gel shown in Fig. 1) was capable of identifying a Dex-induced RNA. Most of the other cDNAs evaluated by North-ern blotting identified bands which were of equal intensity in lanes using RNA from Dex- with vehicle-treated AtT-20 cells. A few cDNAs did not identify any signal on the Northern blot.

The cDNA identified above was cloned into PCR-TRAP and ten clones were partially sequenced. Three unique cDNA species were found. One of the three cDNA clones identified a Dex-induced RNA on a Northern blot (Fig. 2A). This clone was com-pletely sequenced and did not show significant homology with other sequences in GenBank. The cDNA was used as a probe to screen a cDNA library constructed using poly(A)⁺ RNA obtained from Dex-treated AtT-20 cells. Several positive clones were iden-tified and two were sequenced. The full-length gene (1.6 kb) was a new member of the Ras superfamily, the first of this family reported to be induced in response to glucocorticoids (Kemp-painen and Behrend 1998). The gene (called *Dexras1*) is

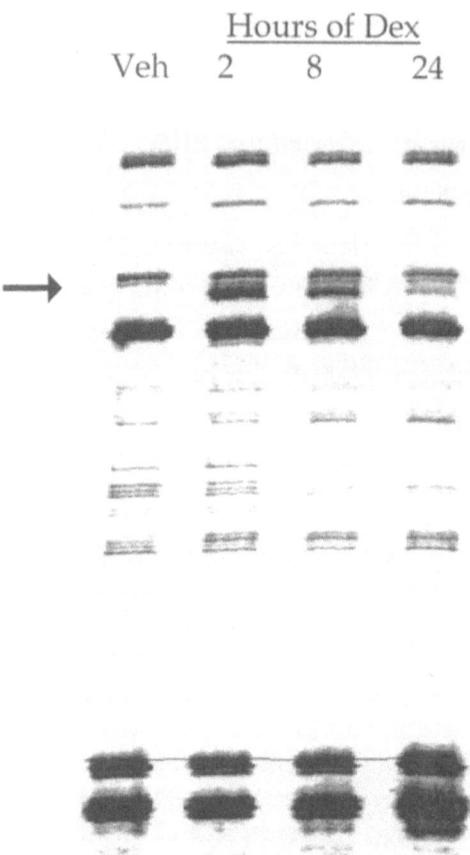

**Fig. 1.** Differential display gel identifying a Dex-induced gene. RNA was obtained from AtT-20 cells treated for 2 h with vehicle (*Veh*) or for 2, 8, or 24 h with Dex (100 nM). Primers for PCR reaction were the one-base anchored primer 5'-AAGCTTTTTTTTTTTC-3' and the arbitrary primer 5'-AAGCTTGTCAGCC-3'. The *arrow* indicates the induced cDNA which was cut from the gel/paper. (From Kemppainen and Behrend 1998 with permission)

expressed in several tissues in mice, and furthermore, expression of the gene is induced in these tissues in response to Dex (Fig. 2B). A time course study of expression in AtT-20 cells showed rapid induction (within 30 min) after treatment with Dex, followed by a steady decline to 24 h post-treatment (Fig. 2C). The role of this gene in glucocorticoid action in general, or negative feedback regulation of ACTH in particular is under investigation.

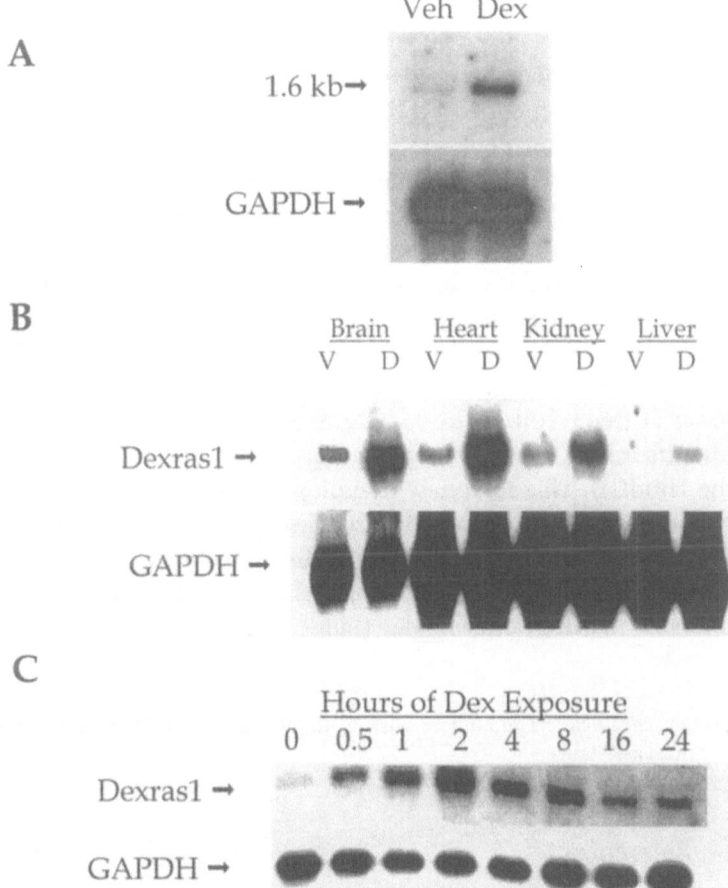

**Fig. 2. A** Northern blot showing that one of the three cDNAs isolated and cloned from the differential display gel in Fig. 1 identified a mRNA induced by Dex in AtT-20 cells. The approximately 260 base pair cDNA was labeled with $^{32}$P by random prime labeling. *Veh* Poly(A)$^+$ RNA (5 µg) from vehicle-treated AtT-20 cells, *Dex* poly(A)$^+$ RNA (5 µg) from AtT-20 cells treated for 2 h with 100 nM Dex. **B** Northern blot using poly(A)$^+$ RNA (5 µg/lane) obtained from tissues in mice treated with vehicle (*V*) or Dex (10 µg/mouse, given intraperitoneally 1 h prior to tissue collection). The probe used was full-length *Dexras1* labeled by the random prime method. **C** Northern blot showing time course of expression of the *Dexras1* gene in AtT-20 cells. Different flasks of cells were treated with 100 nM Dex and were harvested for poly(A)$^+$ RNA at various times after treatment (one flask harvested without any exposure to Dex and represents the 0 h of Dex exposure). The probe used was full-length *Dexras1* labeled by the random prime method; 3 µg poly(A)$^+$ RNA loaded per lane. For loading control, membranes in **A**, **B**, and **C** were stripped of probe and exposed to labeled glyceraldehyde-3-phosphate dehydrogenase (GAPDH) cDNA. (From Kemppainen and Behrend 1998 with permission)

## Comments

Differential display was used to successfully identify a novel gene induced by glucocorticoids in a pituitary endocrine cell line. However, only one gene was identified as being up-regulated by Dex following screening using 240 primer combinations. Other promising cDNAs did not identify induced RNA on Northern blots. This high rate of false positives is commonly reported in association with this technique (Debouck 1995; Zhao et al. 1996). It is probable that other genes are induced in response to Dex in these cells, but we were unable to identify them. Sequence analysis of the *Dexras1* gene showed 100 % homology to the seven bases at the 3' end of the 13-mer arbitrary primer used for its identification (the remaining six bases of the primer comprise the HindIII). This fortunate homology with the 3' bases of the primer probably accounted for the strong signal on the differential display gel (Fig. 1). It should be noted that mismatches between bases on the arbitrary primer and amplified cDNAs are tolerated in differential display (Liang et al. 1995).

*Acknowledgements.* The authors wish to thank Kathleen O'Donnell for her technical assistance.

## References

Ausubel FM, Brent R, Kingston RE, Moore DD, Seidman JG, Smith JA, Struhl K (1995) Current protocols in molecular biology. Wiley, New York

Bertioli DJ, Schlichter UHA, Adams MJ, Burrows PR, Steinbiss H-H, Antoniw JF (1995) An analysis of differential display shows a strong bias towards high copy number mRNAs. Nucleic Acids Res 23:4520–4523

Debouck C (1995) Differential display or differential dismay? Curr Opin Biotech 6:597–599

Kemppainen RJ, Behrend EN (1998) Dexamethasone rapidly induces a novel Ras superfamily member-related gene in AtT-20 cells. J Biol Chem 273:3129–3131

Liang P, Pardee AB (1992) Differential display of eukaryotic messenger RNA by means of the polymerase chain reaction. Science 257:967–971

Liang P, Pardee AB (1997) Differential display methods and protocols. Methods in molecular biology, vol 85. Human Press, Totowa, New Jersey

Liang P, Bauer D, Averboukh L, Warthoe P, Rohrwild M, Muller H, Strauss M, Pardee AB (1995) Analysis of altered gene expression by differential display. Methods Enzymol 254:304–321

Zhang H, Zhang R, Liang P (1996) Differential screening of gene expression difference enriched by differential display. Nucleic Acids Res 24:2454–2455
Zhao S, Ooi, SL, Yang F-C, Pardee AB (1996) Three methods for identification of true positive cloned cDNA fragments in differential display. BioTechniques 20:400–404

## Suppliers

Amersham Life Science, 2636 South Clearbrook Drive, Arlington Heights, Illinois 60005, USA (Tel.: +1-800-3239750, Fax: +1-800-2288735)

Bio-Rad Laboratories, 2000 Alfred Nobel Drive, Hercules, California 94547, USA (Tel.: +1-800-4246723, Fax: +1-800-8792289)

Brinkmann Instruments, P.O. Box 1019, Westbury, New York 11590–007, USA (Tel.: +1-800-6453050, Fax: +1-516-3347506)

Diversified Biotech, 1208 VFW Parkway, Boston, Massachusetts 02132, USA (Tel.: +1-800-7969199, Fax: +1-617-3235641)

Eastman Kodak, 343 State Street, Rochester, New York 14652–4115, USA (Tel.: +1-800-2255352, Fax: +1-800-8794979)

GenHunter Corporation, 624 Grassmere Park Drive, Suite #17, Nashville, 37211, Tennessee (Tel.: +1-615-8330665, Fax: +1-615-8329461)

Life Technologies, P.O. Box 68, Grand Island, New York 14072–0068, USA (Tel.: +1-800-8286686, Fax: +1-800-3312286)

Perkin Elmer Corporation, 850 Lincoln Centre Drive, Foster City, California 94404, USA (Tel.: +1-800-3273002, Fax: +1-415-6385998)

Promega Corporation, 2800 Woods Hollow Road, Madison, Wisconsin 53711–5399, USA (Tel.: +1-800-3569526, Fax: +1-608-2772516)

Qiagen Incorporated, 28159 Avenue Stanford, Santa Clara, California 91355–1106, USA (Tel.: +1-800-4268157, Fax: +1-800-7182056)

Sigma Chemical Company, P.O. Box 14508, St. Louis, Missouri 63178–9916, USA (Tel.: +1-800-3253010, Fax: +1-800-3255052)

Whatman Incorporated, 9 Bridewell Place, Clifton, New Jersey 07014, USA (Tel.: +1-201-7735800, Fax: +1-201-4726949)

# Specialized Cloning
# and Library Screening Strategies

# Two Hybrid cDNA Cloning

JÖRG A. SCHENK

## Introduction

The study of molecular interactions is becoming more and more important in all fields of biomedical and biological research. During the past decade, independent systems for large scale studying of such interactions have been developed and made commercially available.

These systems include phage display and the yeast Two Hybrid System.

The phage display system was originally introduced by G.P. Smith (1985). In this system, a peptide or protein library is expressed on the surface of filamentous phage as fusions to the coat proteins pIII or pVIII. The phage are incubated with the target coated to microtiter plates, columns or magnetic beads. The targets might be proteins, peptides, carbohydrates, haptens, whole cells and even tissues and organs (in vivo panning, Pasqualini and Ruoslahti 1996). Phage carrying specific binders to the target will be selected and non-binders are washed away. The cleaner and more homogeneous the target, the more specific the selected binder(s). Phage display is commonly used to select recombinant antibody fragments from immunized or naive sources (Vaughan et al. 1996) in order to find bioactive peptides. It is also used for epitope mapping (Scott and Smith 1990). There

J.A. Schenk (e-mail: jschenk@mdc-berlin.de, Tel.: +49-30-94062169, Fax: +49-30-94063895)
Max-Delbrück-Center for Molecular Medicine (MDC), Robert-Rössle-Str.10, 13122 Berlin-Buch, Germany
Present address: J.A. Schenk, University of Potsdam, Department of Biotechnology, Golm, Germany

Springer Lab Manual
R.C. Bird, B.F. Smith (Eds.) Genetic
Library Construction and Screening
© Springer-Verlag Berlin Heidelberg 2002

have also been attempts to investigate protein-protein interactions (Choo and Klug 1995; Nord et al. 1997).

The Yeast Two Hybrid System is an assay to detect protein-protein interactions in vivo. It was developed by S. Fields and coworkers (Fields and Song 1989; Chien et al. 1991) and is now commercially available from different companies (e.g. Clontech, Stratagene).

It is useful for rapid identification of genes encoding proteins binding a co-expressed protein target. Many transcriptional activators consist of two distinct modular domains: the DNA-binding domain and the transcriptional activation domain. The DNA binding domain localizes the specific region on the DNA upstream to the transcribed gene and the transcriptional activation domain binds other factors involved in transcription. Usually, both domains are parts of the same protein. Transcription starts only if both domains are present and activated together.

During the 1980s, it was shown that separated domains of one or different transcriptional activators can also be active when assembled in vitro by recombinant DNA technology.

In the Two Hybrid System, such a transcription factor, most commonly yeast GAL4, is separated and both parts are separately cloned as fusion proteins to the proteins of interest. When cotransformed into a yeast strain where the wild-type GAL4 is deleted, such as HF7c or SFY526, only those clones where the fusion proteins interact can transcribe the reporter gene. Reporter genes are usually *lacZ* or *HIS3* so that the final screening can be performed in a color assay or by growing on plates lacking histidine.

The advantage of the system lies in the possibility of also detecting weak and transient interactions. Another characteristic is that in yeast the native conformation of many proteins is more stable than in an in vitro assay. In addition, only the gene or cDNA of a protein and not a purified native protein or antibodies against the protein is required. On the other hand, this also has a serious disadvantage: Only interactions between proteins become detectable, whereas protein–hapten or protein–carbohydrate interactions, as in the phage display, cannot be monitored.

During the past few years, many different interacting protein pairs could be verified or identified using the Two Hybrid System. A list of selected examples is available in: www.fccc.edu/research/labs/golemis/TwoHybridReferences.html.

Today, the Two Hybrid System is an important tool for the identification of the functions of unknown proteins discovered during the large genome sequencing projects.

## Outline

To perform a library screen in the Two Hybrid System, first of all a cDNA-library has to be constructed. The first step is the isolation of total or mRNA from the cells or tissues tested for a binding partner to a known target ("bait") protein. This RNA has to be transcribed into cDNA which subsequently has to be duplicated to form double stranded cDNA. For further cloning, the double-stranded (ds)-cDNA must be ligated to an adaptor with a cohesive end (suggested Eco RI). This end is phosphorylated and the whole cDNA is digested with the restriction endonuclease whose recognition site is situated in the oligo-dT-primer (in this protocol Sal I is suggested). Afterwards the sample is purified and made ready for ligation into the vector containing the GAL4 activation domain (e.g., pGAD424), which has to be digested using the same restriction endonucleases (Eco RI and Sal I) and purified.

As a second prerequisite, the target ("bait") protein's cDNA has to be correctly cloned into a separate vector bearing the GAL4 binding domain (e.g., pGBT9).

The ligated products have to be transformed into bacteria (*E. coli* strains DH5 or XL-1-blue) and amplified. Subsequently, a yeast strain auxotrophic for histidine (e.g., HF7c) has to be cotransformed with both vectors and plated on leucine, tryptophane and histidine-lacking agar plates. Growing clones might be positive and such candidates should then be subjected to further analysis, e.g., beta-galactosidase assay, isolation of plasmid, sequencing or checking by PCR the whole yeast colony for insert screening (Schenk et al. 1996).

A scheme of the procedure is given in Fig. 1.

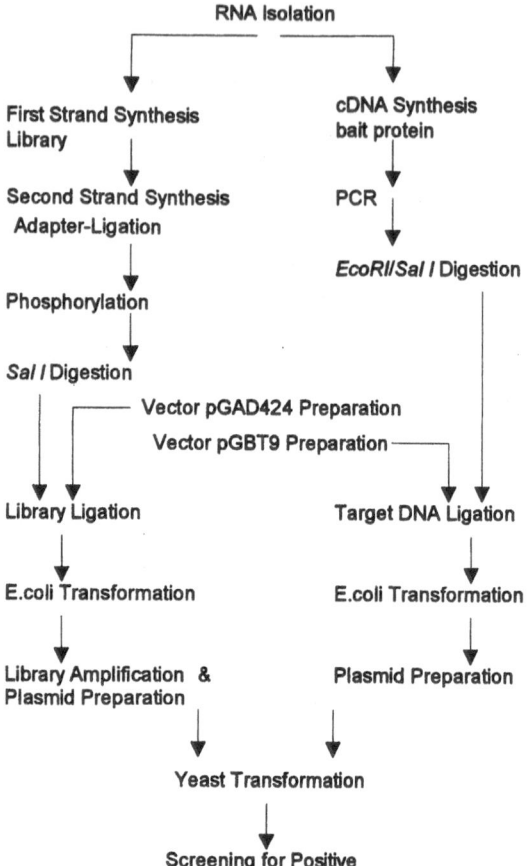

**Fig. 1.** Schematic overview of the procedure

## Materials

The whole cDNA synthesis and cloning procedure might be performed with one of the following kits, where almost all material is included:

–  Two-Hybrid cDNA Library Construction Kit (Clontech, # K 1607–1)
–  HybriZAP-2.1 Two-Hybrid cDNA Gigapack Cloning Kit + HybriZAP-2.1 Two Hybrid cDNA Synthesis Kit (Stratagene, # 235612/#235614)

Note: Materials are only mentioned here if used for the first time.

- Tris-Buffered Phenol
- Chloroform
- Isopropanol
- Glycogen (from mussels, Boehringer Mannheim #901 393)
- Sterile DEPC-treated water
- RNasin (human placental RNase inhibitor, e.g., Boehringer Mannheim #799 017)
- 70 or 75 % ethanol in sterile DEPC-treated water
- RNAzol B (Biotecx Laboratories Inc.) or InViSorb Total RNA Kit II (InViTek GmbH, Berlin-Buch)

**RNA-isolation**

- 10 × Amplification Buffer (160 mM ammonium sulfate, 500 mM Tris-HCl, pH 8.8, 0.1 % Tween 20, supplied with Tth-Polymerase)
- 10 mM $MnCl_2$ (supplied with Tth-Polymerase)
- Tth-Polymerase (e.g., InViScript #10030100, InViTek GmbH, Berlin-Buch)
- 2 mM dNTP (Boehringer Mannheim, #1 277 049)
- 10 pmol/µl Specific Primers
- 100 mM EGTA
- Taq-Polymerase (e.g., InViTAQ DNA Polymerase #10010100, InViTek GmbH, Berlin-Buch)
- Gel-Purifying Kit (InViSorb DNA Extraction Kit #KN2010025, InViTek GmbH, Berlin-Buch; Geneclean #1001–200, Bio 101, Vista, CA; QIAquick Gel Extraction Kit #28704, Qiagen, Hilden)

**Synthesis of bait protein DNA**

- 20 µM Primer 5'-TTGTCGAC(T)$_{25}$(A/G/C)-3'
- 100 mM DTT (Dithiothreitol) (supplied with Revertase)
- 5 × First-strand buffer (250 mM Tris-HCl, pH 8.3, 375 mM potassium chloride, 15 mM magnesium chloride) (supplied with Revertase)
- MMLV reverse transcriptase (Gibco BRL, #28025–013)

**Oligo(dT)-priming**

- 5 × Second-strand buffer (500 mM KCl, 50 mM $NH_4SO_4$, 25 mM $MgCl_2$, 100 mM Tris-HCl (pH 7.5), 0.25 µg/ml BSA)
- 1.5 mM beta-NAD
- RNase H
- DNA-polymerase I (*E. coli*)
- DNA-ligase (*E. coli*)
- T4-DNA-Polymerase

**Second-strand synthesis**

- 0.2 M EDTA
- Phenol-chloroform-isoamylalcohol (25:24:1)
- Chloroform-isoamylalcohol (24:1)
- 4 M NH$_4$-acetate

**Phosphorylation, adaptor-ligation and digestion**

- 10 × One Phor All Buffer PLUS (Pharmacia, #27–0901–02, supplied with Polynucleotidekinase)
- contains: 100 mM Tris-Acetate (pH 7.5), 100 mM magnesium acetate and 500 mM potassium acetate.
- T4-DNA-Ligase (Boehringer Mannheim, #481 220)
- T4-Polynucleotidekinase (Pharmacia, #27–0736–01)
- 100 mM ATP (Pharmacia, #27–2056–01)
- cDNA Spun Columns (Pharmacia, #27–5099–01) or
- SizeSep 400 Spun Columns (Pharmacia, #27–5105–01)
- Eco RI-Adaptors (suggested), synthesize oligonucleotides, the 2nd 5'-phosphorylated and anneal: 5'-AATTCGGCAC-GAG-3', 3'-GCCGTGCTCp-5'
- Alternatively, commercial adaptors may be used, e.g., Eco RI/XhoI-Adapters (Stratagene, #901120)
- Restriction Endonuclease Sal I (New England Biolabs, #138S)

**Vector preparation and library ligation**

- Vectors pGBT9 and pGAD424 (Clontech, #K 1605-A and #K 1605-B) or MATCHMAKER Two-Hybrid System (Clontech, #K 1605-1)
- Restriction endonuclease Eco RI (New England Biolabs, #101S)
- Agarose (Sigma, #A-9539)
- LB broth (1 liter contains 10 g Bacto-Tryptone, 5 g yeast extract and 10 g NaCl)
- Electrocompetent DH5α or XL-1-Blue Cells (Clontech, #C2022–1, Stratagene, #200228)
- Electroporation cuvettes 0.2 cm (Biorad, Eurogentech)
- Electroporator
- Carbenicillin (Sigma, #C 1389)
- Qiagen Plasmid Maxi Kit (Qiagen, #12162)

# Procedure

The following procedure is only one of several established protocols. All steps might be modified according to the experience and
predilections of the respective researcher.

### RNA isolation

Depending on the amount of starting material the appropriate
isolation method should be selected. There are two general methods of isolating total RNA: the acid guanidinium thiocyanate-
phenol procedure (AGCP) (Chomcynski and Sacchi 1987) and
the carrier method (Vogelstein and Gillespie 1979; Schenk et al.
1997). The carrier method is more suitable for small amounts of
starting material, while for library construction, the AGCP might
be better. For all RNA isolation procedures, sterile gloves should
be worn to avoid skin contact with the solutions and materials
involved in RNA isolation. All solutions used, including ethanol
dilutions, should be prepared using sterile DEPC-treated RNase-
free double distilled water. To avoid RNase contamination, it is
recommended to carry out RNA isolation only in labs where no
bacterial, yeast or cell cultures are grown.

1. Add 200 µl RNAzol to $10^6$ sedimented cells. If cells are grown **RNA isolation**
   in a monolayer, add RNAzol directly to the culture dish (1 ml **using RNAzol**
   for each 3.5-cm petri dish).

2. Solubilize RNA by pipetting a few times up and down.

3. Add 100 µl chloroform per ml cells, vortex for 15 s and place
   on ice for 5 min.

4. Centrifuge at 12,000 rpm for 15 min at 4 °C.

5. Carefully transfer the **upper** aqueous phase to a fresh tube.

6. Add an equal volume of isopropanol and 1 µl glycogen.

7. Place on ice for 15 min. (This step can be extended if necessary. It is possible to store the tubes for several weeks at
   –20 °C).

8. Centrifuge at 12,000 rpm for 15 min. at 4 °C. A white-yellow pellet (RNA) should be visible.

9. Remove supernatant and add about 1 ml 75% ethanol (in DEPC-treated water) for 50–100 µg RNA. Vortex and centrifuge at 7500 rpm for 10 min at 4 °C.

10. Dry the pellet briefly under vacuum, but do not let it dry completely.

11. Dissolve the pellet in sterile DEPC-treated water. Sometimes it might be necessary to incubate the tube at 60 °C for several minutes to dissolve RNA completely.

12. For storage add 1 µl of human placental RNase inhibitor.

**RNA-isolation using InViSorb Total RNA Kit II**

1. Spin down about $10^6$ cells gently and wash twice with PBS. Add 500 µl of Chaotropic Lysis Buffer. If cells are grown as a monolayer in cell dishes, lysis can be performed directly by adding 500 µl Chaotropic Lysis Buffer.

2. Add 30 µl Adsorbin (mineral carrier) solution and vortex thoroughly.

3. Place the tubes on ice for 5 min to allow DNA binding to the carrier.

4. Centrifuge for 1 min at 15,000 rpm to pellet the adsorbin.

5. Gently transfer the supernatant to a fresh microcentrifuge tube.

6. Add 500 µl tris-buffered phenol, 100 µl chloroform and 50 µl buffer A, vortex vigorously for 15 s and place the tube on ice for 5 min.

7. Centrifuge for 10 min at 15,000 rpm and 4 °C.

8. Gently transfer the upper aqueous phase into a fresh microcentrifuge tube. It is better to leave a few microliters of the aqueous phase than contaminate it with the organic or interphase!

9. Add an equal volume of isopropanol and 1 µl glycogen, mix and incubate at −20 °C for 20–30 min.

10. Centrifuge at 15,000 rpm and 4 °C for 10 min. An RNA pellet should be visible.

11. Carefully remove supernatant and add 1 ml ice-cold 70 % ethanol.

12. Mix gently by pipetting up and down several times and centrifuge for 3 min at 15,000 rpm and 4 °C. Remove ethanol and wash at least once more.

13. Dry the RNA pellet, but do not let it dry completely.

14. Dissolve the RNA pellet in 50–100 µl (depending on the initial cell number) DEPC-treated water or TE-buffer. If RNA dissolves badly, incubate the tube for 2 min at 60 °C.

Optionally, mRNA may be purified from the isolated total RNA using the mRNA Separator Kit (Clontech # K1040-2)

### cDNA synthesis

After isolation the RNA has to be transcribed into DNA for further cloning. Long-time storage of samples is much more convenient after cDNA synthesis than in RNA form due to higher stability of cDNA. cDNA synthesis using RNA templates is carried out using the enzyme reverse transcriptase or revertase. There are three different methods to prime the synthesis. Most common for library construction is the priming with oligo(dT). Since mRNA usually carries a poly-A-tail, an oligomer of dT with a length of 14–25 nucleotides as a primer will selectively bind to mRNA rather than to rRNA or tRNA. Thus after cDNA synthesis using total RNA, only mRNA will be transcribed into cDNA. For use with oligo(dT)-primers, a viral reverse transcriptase (usually from AMV or MMLV) is recommended because the reaction temperature (42 °C) is in the range of the annealing temperature (ca. 28–50 °C). It has been shown that MMLV reverse transcriptase produces consistently better yields of cDNA than AMV reverse transcriptase (Chenchik et al. 1994). To prime exclusively at the end of the poly(A)-tail, primers containing $(dT)_{14-25}(dA/G/C)$ are used. As further cloning of this cDNA is needed, insertion of a restriction site into the tail of the $(dT)_n$-primer is recommended (Sal I in this protocol).

Another method of priming is the use of hexa-random-oligonucleotides. Using this method, all RNA is transcribed into cDNA. Viral revertase is also recommended for such a reaction. This is only recommended for library construction if the RNA has been isolated with a specific method for mRNA isolation (poly-T column, etc.). Otherwise the background of transcribed tRNA and rRNA will disturb real mRNA constructs as they are only about 1–2 % of total RNA.

If the RNA that is to be transcribed is well known, the use of sequence specific primers as in a further PCR reaction is a good choice. In this case the background can be minimized since only specific mRNA is transcribed into cDNA. Normally the annealing temperature of such primers is higher than 42 °C and viral revertases are inactive at such temperatures. That is why a bacterial polymerase, Tth-Polymerase of *Thermus thermophilus*, should be used for such a reaction. This enzyme usually catalyzes the synthesis of DNA from a DNA template in the presence of magnesium ions. If these ions are replaced by manganese ions, the activity switches to synthesis of DNA from an RNA template.

The specific primers should be selected with 5'-overhangs containing restriction sites also present in the respective Two Hybrid Vector. Make sure that cutting these sites leaves the gene in the correct reading frame and that the restriction sites are not present in the gene! Remember that restriction endonucleases usually do not cut if their recognition sequence is too near to the end of the primer (for details see New England Biolabs Catalog, 1996/97, pp. 238/39 for details).

**Synthesis of the bait protein DNA**

1. Mix 10 µl (1/2 isolate of $10^5$ cells) total RNA with 2 µl 10 × amplification buffer, 2 µl 10 mM $MnCl_2$, 1 µl human placental RNasin, 2 µl 2 mM dNTP-mix, 20 pmol specific back primer and 1 µl Tth-polymerase (InViScript) in a final volume of 20 µl on ice.

2. Spin down and heat at 65 °C for 5 min.

3. Incubate 2 min at 30 °C.

4. Incubate 20 min at 70 °C.

5. Stop the reaction by adding 0.4 µl 100 mM EGTA.

6. Mix 1 µl of the cDNA mixture from step 5 with 10 µl 10 × amplification buffer, 5 µl 50 mM $MgCl_2$, 10 µl 2 mM dNTP-mix, 50 pmol of each primer (forward and backward) and 4 units Taq-Polymerase in a final volume of 100 µl.

7. Spin down and perform PCR for about 30 cycles. The PCR conditions must be evaluated for each gene separately.

8. Analyze 5 µl of the PCR reaction product on a 1.5 % agarose gel. If the expected band is present purify remaining 95 µl of the product using a commercial PCR Product Purifying Kit

9. Digest the purified PCR product with the restriction endonucleases present in the primers. Perform digestion for at least 6 h or overnight (usually at 37 °C depending on the enzymes).

10. Run another 1.5 % agarose gel and excise the band according to the digested product. Purify the DNA from the gel with a Gel Extraction Kit.

11. The gene of the bait protein is now ready for cloning.

**cDNA-synthesis using oligo(dT)-primers**

1. Mix 3–5 µg of isolated Poly(A)-RNA with 5 µl of Oligo(dT)25(dA/C/G)-Sal I primers (20 µM).

2. Add 1 µl RNasin (Human Placental RNase Inhibitor).

3. Add sterile DEPC-treated water to a final volume of 12.5 µl.

4. Heat the tube in a 70 °C water bath for 3 min to destroy RNA secondary structure.

5. Cool down in an ice bath for several minutes to allow primers to anneal. Spin the tubes briefly in a microcentrifuge.

6. Add the following components to the reaction tube: 5 µl 5 × first-strand buffer, 1.3 µl 20 mM dNTP, 0.5 µl DTT (100 mM), 3.2 µl sterile DEPC-treated water. Mix gently.

7. Add 500 units (2.5 µl) MMLV reverse transcriptase. Mix gently and spin the tubes in a microcentrifuge.

8. Incubate at 42 °C for 90 min.

9. Place the tubes on ice and proceed with second-strand synthesis.

**Second-strand synthesis**

1. Add to the tube 40 µl 5 × second-strand buffer, 1.5 µl 20 mM dNTP, 20 µl 1.5 mM β -NAD, 2.5 units RNaseH, 60 units DNA-polymerase I (*E. coli*) and 12 units DNA-ligase (*E. coli*) and sterile DEPC-treated water to a final volume of 200 µl.

2. Incubate at 16 °C for 120 min.

3. Add 15 units T4 DNA polymerase and incubate for another 30 min at 16 °C.

4. The reaction is stopped by adding 10 µl 0.2 M EDTA.

5. Purification of ds-cDNA is necessary now. You may use a commercially available kit or proceed with step 6.

6. Add 200 µl phenol/chloroform/isoamyl alcohol to the tube and mix for 2 min.

7. Centrifuge the sample for 2 min at 15,000 g.

8. Transfer the aqueous upper phase to a fresh microcentrifuge tube.

9. Add 200 µl chloroform/isoamyl alcohol and mix for 2 min, repeat steps 7 and 8

10. Add 200 µl 4 M $NH_4$-acetate and 1 ml ice-cold 96 % ethanol.

11. Mix gently and place the tube in the freezer (-20 °C) for 1 h or in an ethanol/dry-ice bath (-70 °C) for 10 min.

12. Centrifuge the sample for 20 min at 15,000 g.

13. Carefully remove the supernatant and air dry the pellet for about 15 min or under vacuum for 2 min.

### cDNA preparation and cloning

**Adapter ligation**  The double stranded cDNA constructed in the last step has to be cloned into the respective Two-Hybrid vectors to perform the screening. As the oligo-(dT)-primed cDNA has only one incorporated restriction site, another site for cloning has to be inserted now. Therefore, oligonucleotides already containing the base overhangs for cloning, are ligated to the ends of the ds-cDNA. As the sequence of the generated DNA is usually unknown, digestion with restriction endonucleases could lead to undesired internal cuts. So this risk is reduced twofold.

1. Dissolve the pellet in 15 µl sterile DEPC-treated water.

2. Add 5 µl 10 × One Phor All-Buffer *PLUS*, 5 µl 10 mM ATP, 10 units T4-DNA-ligase, 4 µg Eco RI-adapters and sterile DEPC-treated water to a final volume of 50 µl.

3. Mix gently and incubate overnight at 16 °C.

4. Increase the temperature to 22 °C and incubate for 4 h more.

5. Inactivate the ligase at 70 °C for 30 min. Spin in a microcentrifuge for some seconds and store the sample on ice.

1. Add 1 µl 10 × One Phor All-Buffer *PLUS*, 2 µl 10 mM ATP, 6 µl sterile water and 10 units (1 µl) T4-polynucleotide kinase **Phosphorylation of the adapter ends**

2. Incubate at 37 °C for 30 min.

3. Inactivate the Polynucleotide Kinase for 30 min at 70 °C.

4. Spin the sample in a microcentrifuge for several seconds and equilibrate at room temperature for some minutes.

1. Add 4 µl One Phor All-Buffer *PLUS* and 100 units Sal I in a final volume of 100 µl. **Digestion with Sal I**

2. Incubate at 37 °C for 90 min.

3. Inactivate the enzyme by incubation at 65 °C for 20 min.

4. To remove very short cDNAs (<300 bp), unligated adaptors, digested fragments and remaining nucleotides, the cDNA has to be purified using cDNA spun columns. Such columns are available from several companies, (e.g., Pharmacia, Clontech) which have different protocols provided with the columns.

5. Precipitate purified cDNA fractions by adding 1.5 µl glycogen, 1/2 volume 7.5 M $NH_4$-acetate and 2.5 volumes 96 % ethanol. Place in an ethanol/dry ice bath for at least 2 h or in a –70 °C freezer overnight for a better result.

6. Centrifuge the sample in a microcentrifuge for at least 20 min at 15,000 rpm.

7. Carefully remove the supernatant and dry the pellet.

8. Resuspend the pellet in 6 µl TE buffer. The cDNA is now ready for ligating into the Two-Hybrid vector.

1. Add to 10 µg vector (containing the GAL4-activation domain, e.g., pGAD424) 10 µl 10 × One Phor All Buffer *PLUS*, 10 units Sal I, 10 units EcoRI and sterile water to a final volume of 100 µl. **Vector preparation**

2.  Incubate at 37 °C for 3 h.

3.  Run a 1 % agarose gel, stained with ethidium bromide, and excise the band corresponding to the length of double digested vector.

4.  Purify the digested vector from the gel with a DNA purifying kit (e.g., Geneclean, Qiagen, InViTek, for details see above "Synthesis of bait protein DNA"). Dissolve the vector finally in a volume of 20 μl TE buffer or water.

5.  Repeat the procedure for the vector containing the GAL4-binding domain, e.g., pGBT9 and the restriction endonucleases whose recognition sites are present in the specific primers, e.g., Sal I and Eco RI could have been used here.

**Ligation of cDNA into the vector**

To find out the optimal ratio of vector to cDNA-insert, different ratios are tested. One ligation reaction is carried out containing no insert as a control. The number of colonies from the library ligation should be at least 100 times larger than that of the control.

1.  Mix in a microcentrifuge tube the following reagents for ligation:

    5 μl One Phor All Buffer *PLUS*, 5 μl 10 mM ATP, 100 units T4-DNA-ligase, 1 μl vector (pGAD424 for library ligation), 3, 2, 1, and 0 μl purified and digested ds-cDNA, respectively, in a final volume of 50 μl (add sterile water). So there are four different tubes for ligation.

2.  Incubate at 16 °C for at least 4 h or overnight.

3.  Add 1 μl glycogen and 200 μl ice-cold 96 % ethanol. Mix well.

4.  Place in an ethanol/dry-ice bath for 2 h or overnight.

5.  Centrifuge at 15,000 rpm for 20 min.

6.  Carefully remove supernatant and air-dry the pellet .

7.  Dissolve the pellet in 5 μl sterile distilled water.

8.  Repeat steps 1–7 using vector pGBT9 and digested and purified bait-protein DNA. Both ligations might be carried out in parallel as well.

## Transformation of bacteria

The ligated plasmids with the cDNA library and the bait protein DNA have to be amplified in *E. coli* (usually strains DH5 or XL-1-Blue) before transfecting yeast. There are different methods of bacterial transformation, but usually electroporation has the highest yields. Transformation efficiency plays an important role in the general success of the Two Hybrid experiment.

Electrocompetent cells are available from several companies but it is also possible to prepare them by yourself (Potter 1993). It **is very important to store cells on ice at all steps until broth is added.**

1. Add the dissolved ligation product to 100 µl electrocompetent cells, mix gently and wait for 1 min.

2. Transfer the cells to a 2-mm cuvette and pulse with the electroporator with the following parameters: 2.5 kV, 25 µF, 200 Ohms. In the case of arching, add another 100 µl cells to the reaction and try again.

3. Immediately add 900 µl prewarmed (37 °C) LB or SOC broth to cuvette and shake gently.

4. Transfer the solution to a 5-ml tube and shake for 60 min at 37 °C.

5. Transfer 10 µl of each transformation into 40 µl LB broth and plate onto 1.5 % agar plates containing 50 µg/ml carbenicillin. Incubate plates at 37 °C overnight.

6. Store the remaining transformation samples overnight at 4 °C.

Next morning count the colonies and determine the best vector/insert ratio. If the cDNA ligations have at least 100 times more colonies than the control plate with vector alone, the ligation was successful. Pool the stored samples of the positive transformations and store the library until further use at 4 °C.

Pick some colonies (if the ratio vector-alone/vector-plus-insert was good, ten should be sufficient) from the bait-protein ligation and perform a PCR with the whole colonies and the specific primers for the bait-protein's DNA or the pGBT vector primers (Schenk et al. 1996).

1. Mix 10 µl 10 × amplification buffer, 5 µl 50 mM MgCl$_2$, 10 µl 2 mM dNTP-Mix, 2 units Combi-Pol-polymerase, 50 pmol of each primer in a final volume of 100 µl.

2. Preheat the mixture to 94 °C.

3. Take a pipette tip, scratch a single bacterial colony from the bait protein-ligation plate, tip it onto a fresh agar plate containing 50 µg/ml carbenicillin (reference plate) and afterwards into the preheated reaction tube. Mix gently.

4. Run a PCR with the appropriate conditions for the respective bait protein gene. 30 cycles should be enough to detect whether an insert is present or not.

5. Analyze 5 µl of each reaction on a 1 % agarose gel stained with ethidium bromide.

6. Incubate the reference plate overnight at 37 °C.

7. Inoculate a positive colony into 200 ml LB broth containing 50 µg/ml carbenicillin and grow overnight in a shaking incubator at 37 °C and 250 rpm.

8. Isolate and purify plasmid according to Sambrook et. al.(1989) or with a kit (Qiagen Plasmid Maxi Kit). Determine concentration by UV analysis.

9. Check an aliquot of the isolated plasmid by double digestion with Eco RI and Sal I (or the respective cloning enzymes) and run a 1 % agarose gel. There should be two bands: One should correspond to the length of the insert DNA and the other to the length of vector pGBT9 (5400 bp).

10. The pGBT9 vector is now ready for yeast transformation.

### Amplification of the cDNA library

To acquire a sufficient quantity of pGAD424 plasmid for yeast transformation, the cloned cDNA library has to be amplified. Growth in liquid culture overnight could lead to unequal amplification of different clones and the library would not be representative anymore. For that reason plating on LB-agar plates containing 50 µg/ml carbenicillin is recommended.

The plating of the aliquots results in an estimated number of clones for the whole library. For example: If one of the ligation mixtures results in 1000 colonies on the control plate, there should be about 100,000 different clones from this ligation mixture (10 µl out of 1 ml).

Calculate how many agar plates are needed. Usually there should be 10,000 to 20,000 clones per plate. Divide the volume of the library by the calculated number of plates to get the volume for each plate (it should be at least 50 µl).

1. Spread the stored library onto LB-agar plates (150 mm) containing 50 µg/ml carbenicillin.

2. Incubate the plates overnight at 37 °C.

3. Add 5 ml LB broth to one plate. Scrape the colonies off the agar surface and transfer the liquid to a sterile, prechilled vessel.

4. Repeat step 3 until all plates are scraped. Store the vessel on ice while scraping.

5. Take a desired amount of the "harvested" library (100 ml, depending on the estimated complexity) and isolate the plasmid (according to Sambrook et al. 1989, or with the Qiagen Plasmid Maxi Kit). Add sterile glycerol at a final concentration of 25 % to the remaining library and freeze it in sterile tubes and small aliquots (1 ml each) at –80 °C.

6. Determine concentration by UV.

7. A control digestion of a small aliquot of the isolated plasmid library using Eco RI and Sal I is recommended. There should be a strong band corresponding to the length of plasmid pGAD424 and a smear of other bands corresponding to the lengths of the different cloned cDNAs.

8. The pGAD424-cDNA library is now ready for yeast transformation.

Now, both plasmids for yeast transformation are present in sufficient amounts. There are different methods for yeast transformation: electroporation and several chemical procedures. Which method should be applied depends on the equipment and experience of the laboratory. Furthermore, both plasmids might be cotransformed or the pGBT9 might be transformed first and positive colonies might be transformed with the pGAD424 plasmid in a second round.

The Clontech Two Hybrid System Manual and the publication by Cowell (1997) include protocols of how to perform the yeast transformation, yeast screening and identification of positive colonies. Common yeast protocols are also available from Sherman (1991).

## Troubleshooting

- Poor RNA
    1. RNA is very sensitive to degradation by RNases in the biological material used for RNA extraction. Therefore, it is important to use only the freshest material available. Whenever possible, RNA isolation should be started immediately after collection of the biological material.
    2. If you suspect a failure of RNA isolation, a denaturing formaldehyde/agarose gel should be performed. Total RNA should appear as a smear from 0.5–18 kb with very bright bands at about 4.5 and 1.9 kb corresponding to 28S and 18S rRNA. The ratio of these bands should be higher than 1.5:1.
    3. The amount of isolated RNA can be quantified at 260 nm. It is calculated with the formula: 0.1 OD unit × dilution factor (usually 200–1000, depending on the expected amount) × 40 µg/ml=1 µg/µl.

- Poor cDNA synthesis
    1. To monitor the cDNA synthesis, traces of radioactive nucleotides (1 µl of $\alpha^{32}$P-dCTP) might be added to the first strand reaction (800 Ci/mmol). After precipitation of the cDNA pellet, the radioactivity of the pellet should be at least 100,000 cpm.
    2. The cDNA might be examined on an alkaline gel according to Feilotter et al. (1994). It should appear as a smear ranging from 0.5–10 kb.
    3. Make sure that enzymes are added immediately before use. Vortex the buffers vigorously.

4. Incubation temperatures higher than 16 °C during the second strand synthesis might also lead to insufficient results.

5. For all precipitation steps make sure that 96 % ethanol is used. Use of 70 % ethanol might lead to loss of the pellet.

- Poor ligation, digestion and transformation

1. Use of very high doses of enzymes leads to an excess of glycerol in the reaction mixture. Usually, the final glycerol concentration must be less than 5 %. Make sure that no additional enzyme adheres to the outside of the pipette tip.

2. Ensure that ATP is added to the ligation reaction.

3. Make sure that the chosen restriction endonuclease is able to digest restriction sites near the end of the DNA.

4. If the numbers of colonies are low, check the competent cells by transformation of a supercoiled control plasmid. Make sure that commercial competent cells are not thawed during shipping!

5. If the ratio ligation with/without-insert is bad, try other vector/insert ratios and make sure that both, vector and insert have the correct cohesive ends.

*Acknowledgements.* The author would like to thank Prof. B. Micheel and Dr. S. Heymann for their critical review of the manuscript and Mrs. E. Micheel for correcting the English manuscript.

# References

Chenchik A, Diatchenko L, Chang C, Kuchibhatla S (1994) Great lengths cDNA synthesis kit for high yields of full-length cDNA. Clontechniques 9(1):9–12

Chien CT, Bartel PL, Sternglanz R, Fields S (1991) The Two-hybrid system: a method to identify and clone genes for proteins that interact with a protein of interest. Proc Natl Acad Sci USA 88:9578–9582

Chomczynski P, Sacchi N (1987) Single step method of RNA isolation by acid guanidinium thiocyanate-phenol-chloroform extraction. Anal Biochem 162:156–159

Choo Y, Klug A (1995) Designing DNA-binding proteins on the surface of filamentous phage. Curr Opin Biotechnol 6:431–436

Cowell IG (1997) Yeast two-hybrid library screening. In: Cowell IG, Austin CA (eds) cDNA library protocols. Methods in molecular biology, vol 69. Humana Press, Totowa, NJ, pp 185–202

Feilotter HE, Hannon GJ, Ruddell CJ, Beach D (1994) Construction of an improved host strain for two-hybrid screening. Nucleic Acids Res 22:1502–1503

Fields S, Song O (1989) A novel genetic system to detect protein-protein interactions. Nature 340:245–247

Nord K, Gunneriusson E, Ringdahl J, Stahl S, Uhlen M, Nygren PA (1997) Binding proteins selected from combinatorial libraries of an alpha-helical bacterial receptor domain. Nat Biotechnol 15:772–777

Pasqualini R, Ruoslahti E (1996) Organ targeting in vivo using phage display peptide libraries. Nature 380:364–366

Potter H (1993) Application of electroporation in recombinant DNA technology. Methods Enzymol 217:461–483

Sambrook J, Fritsch EF, Maniatis T (1989) Molecular cloning: a laboratory manual, 2nd edn. Cold Spring Harbor Laboratory Press, Cold Spring Harbor, NY

Schenk JA, Heymann S, Peters LE, Micheel B (1996) Screening for recombinant plasmids in yeast colonies of the two hybrid system using PCR. Biotechniques 20:812–816

Schenk JA, Hillebrand T, Lübbe L, Heymann S, Böttger M, Micheel B, Bendzko P (1997) Fast isolation of RNA to detect expression of tumor markers. J Clin Lab Anal 11:340–342

Scott JK, Smith GP (1990) Searching for peptide ligands with an epitope library. Science 249:386–90

Sherman F (1991) Getting started with yeast. Methods Enzymol 194:3–21

Smith GP (1985) Filamentous fusion phage: novel expression vectors that display cloned antigens on the virion surface. Science 228:1315–1317

Vaughan TJ, Williams AJ, Pritchard K, Osbourn JK, Pope AR, Earnshaw JC, McCafferty J, Hodits RA, Wilton J, Johnson KS (1996) Human antibodies with sub-nanomolar affinities isolated from a large non-immunized phage display library. Nat Biotechnol 14:309–314

Vogelstein B, Gillespie D (1979) Preparative and analytical purification of DNA from agarose. Proc Natl Acad Sci USA 76: 615–619

## ▨ Suppliers

Roche Diagnostics GmbH, Sandhofer Strasse 116, 68305 Mannheim, Germany (e-mail: mannheim.biocheminfo@roche.com, USA, Fax: +49-621-7598509, Germany)

Clontech Laboratories Inc., 1020 East Meadow Circle, Palo Alto 94303–4230, California, USA (e-mail: orders@clontech.com, Fax: +1-800-4241350)

GIBCO BRL Life Technologies Ltd., 3 Fountain Drive, Inchinnan
Business Park, Paisley PA4 9RF, UK (Fax: +44-800-243485)

InViTek GmbH, Robert-Rössle-Strasse 10, 13125 Berlin-Buch,
Germany, http://www.invitek.de
Fax: +49-30-94893795)

New England Biolabs, Inc., 32 Tozer Road, Beverly 01915–5599,
Massachusetts, USA (e-mail: info@neb.com,
Fax: +1-508-9211350)

Qiagen GmbH, Max-Volmer-Strasse 4, 40724 Hilden, Germany
(Fax: +49-2103-892233)

Qiagen Inc., 28159 Avenue Stanford, Santa Clara, California
91355, USA (Fax: +1-800-7182056)

Sigma, P.O. Box 14508, St.Louis, Missouri 63178–9916, USA
(Fax: +1-800-3255052, http://www.sigmaaldrich.com)

Stratagene Ltd. (e-mail: paul@stratagene.co.uk,
Fax: +44-1223-420234, UK, +1-800-4245444, USA)

# Yeast One and Two Hybrid cDNA Cloning

BETTY C.B. HUANG and YING LUO

## ▨ Introduction

To understand the functions of a protein, it often requires
cloning of the other cellular proteins that interact with it. Several
approaches have been used by biologists to clone new genes
based on protein-protein interactions, including co-immunopre-
cipitation, affinity purification, and the yeast two-hybrid system.
The yeast two-hybrid cDNA cloning technology is a powerful in
vivo protein-protein interaction assay first introduced in 1989 by
Fields et al. (Fields and Song 1989; Chien et al. 1991; Chevray and
Nathans 1992; Durfee et al. 1993; Zervos et al. 1993; Mendelsohn
and Brent 1994). It is based on co-expression of two proteins, X
and Y, fused to GAL4 DNA binding domain (GAL4B) and GAL4
transcription activation domain (GAL4A) respectively (Fig. 1A).
If the protein X interacts with the protein Y, the GAL4 transcription activation domain will be brought to the promoter containing the GAL4 DNA binding sites and will activate the transcription of reporter gene HIS3 or lacZ. The two-hybrid system can be
used to clone cDNA encoding a novel protein that interacts with
a known protein (bait) in yeast. It can also be used to study protein-protein interactions between two known proteins. The yeast
two-hybrid system has several advantages over other conventional methods used to study protein-protein interactions:

B.C.B. Huang (e-mail: bcbhuang@rigel.com), Y. Luo (e-mail:
yluo@rigelinc.com, Tel.: +1-650-6241130, Fax: +1-650-6241133)
Rigel Inc., 240 East Brand Avenue, South San Francisco, California 94080,
USA

Springer Lab Manual
R.C. Bird, B.F. Smith (Eds.) Genetic
Library Construction and Screening
© Springer-Verlag Berlin Heidelberg 2002

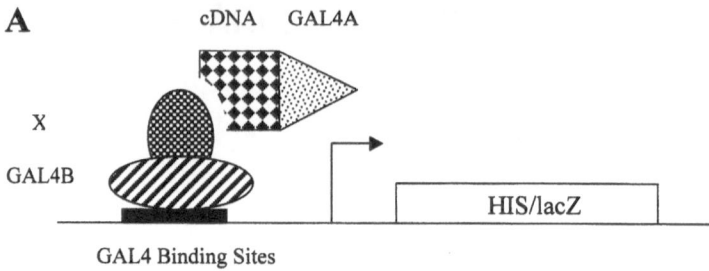

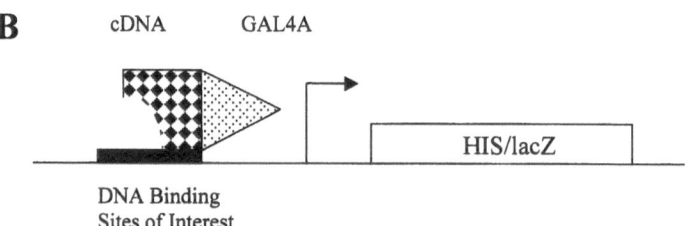

**Fig. 1.** Underlying mechanism of yeast two-hybrid and yeast one-hybrid systems. A Two-Hybrid System: *GAL4A* represent GAL4 transcription activation domain. *cDNA* represents cDNA library inserts. *X* represents any bait gene. *GAL4B* represents GAL4 DNA binding domain. *HIS/lacZ* indicates that the reporter gene is either HIS or lacZ

- Protein-protein interactions are studied in eukaryotic cells (yeast).

- cDNA clones are immediately available after screening.

- Screening is very fast and convenient. No protein purification is necessary.

- Both growth selection and lacZ color selection are very sensitive.

- No radioisotope is needed.

However, due to certain technical difficulties, this technology also presents a challenge to new users:

- Depending on the bait protein used in screening, false positive rates vary significantly.

- Yeast transformation efficiency is difficult to control.

- High quality cDNA libraries can be very difficult to construct.

The major difference between the yeast one-hybrid system and the yeast-two hybrid system is that the one-hybrid system is used to clone cDNA encoding DNA-bound proteins, rather than protein-bound proteins (Fig 1B). DNA sequences of interest are inserted upstream of the minimal promoter controlling the expression of either HIS or lacZ gene. cDNA fragments are fused to the C-terminal of GAL4 transcription activation domain (GAL4A) to construct the cDNA library. If the protein encoded by the cDNA can bind to the specific DNA sequences of interest, the transcription of HIS/lacZ is activated. The yeast one-hybrid system is widely used in cloning new transcription factors (Wilson et al. 1991; Li and Herskowitz 1993; Wang and Reed 1993; Lehming et al. 1994; Strubin et al. 1995; Luo et al. 1996; Shang et al. 1997). Experimental protocols between these two methods are very similar except for one notable exception. The one-hybrid system yeast reporter strains need to be constructed by the individual researcher. The expression of reporter genes (HIS/lacZ) should be under the control of specific DNA sequences of interest, rather than the GAL4 DNA-binding sites in the yeast two-hybrid system. cDNA libraries used for two-hybrid screening can also be used for one-hybrid screening.

## Outline

The experimental flow-chart of a yeast two-hybrid cDNA screening experiment is outlined in Fig. 2. The experimental flow-chart of a yeast one-hybrid cDNA screening experiment is outlined in Fig. 3.

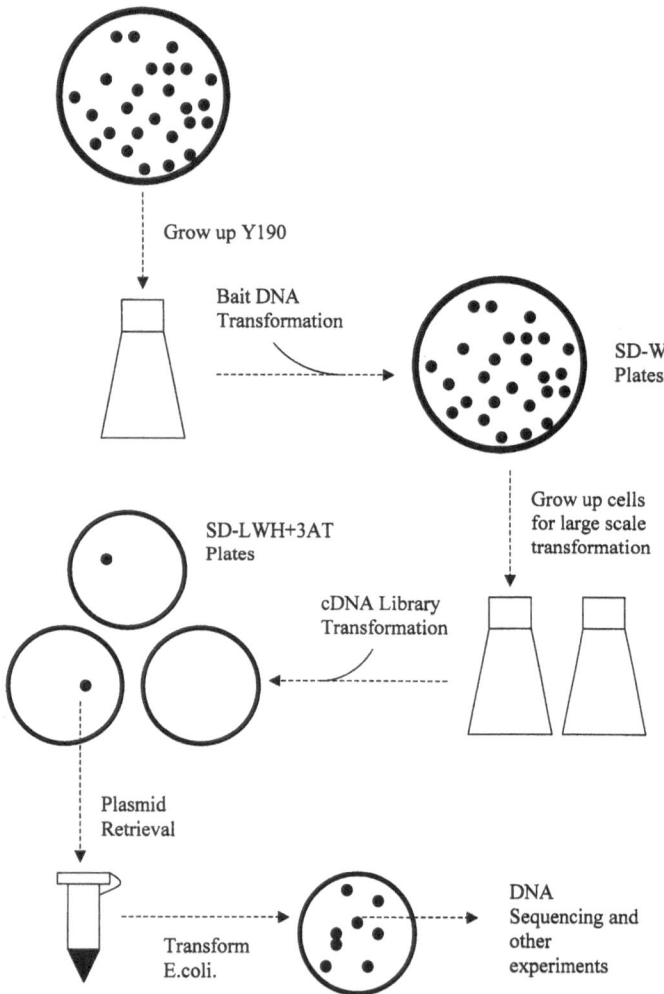

**Fig. 2.** Outline of Yeast Two-Hybrid Screening. The entire process takes about 3 weeks. *Solid black dots* represent colonies on plates. Transformation steps of both bait plasmid and cDNA library plasmids are indicated

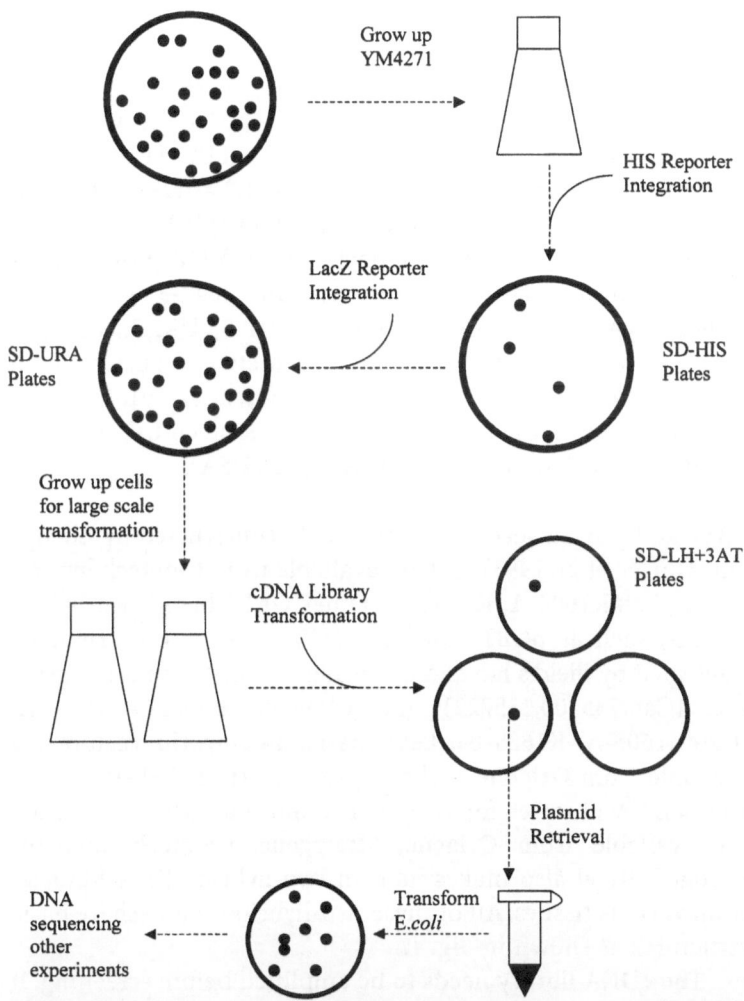

**Fig. 3.** Outline of Yeast One-Hybrid Screening. The entire process takes about 3 weeks. The testing step for optimal 3AT concentration is not included. *Solid black dots* represent colonies on plates

## ▪ Materials

**Medium and yeast strains**

All yeast culture mediums, including YPD, YPD Agar, DOB, DOBA, CSM-TRP, CSM-LEU, CSM-HIS, CSM-URA, CSM-LYS, CSM-LEU-TRP, CSM-LEU-HIS, and CSM-LEU-TRP-HIS, are available from Bio101, Inc. 3AT (3-amino-1,2,4-triazol) is available from Sigma (Cat# A-8056, St. Louis, MO, USA).

Yeast two-hybrid system reporter strain Y190 (**MATa**, ura3-52, his3-200, lys2-801, ade2-101, trp1-901, leu2-3, 112, gal4Δ, gal80Δ, cyhr2, LYS2::GAL1$_{UAS}$-HIS3$_{TATA}$-HIS3, URA3::GAL1$_{UAS}$-GAL1$_{TATA}$-lacZ) and yeast one-hybrid system reporter strain YM4271 (**MATa**, ura3-52, his3-200, lys2-801, ade2-101, trp1-903, leu2-3, 112, tyr1-501) are available from Clontech Laboratories, Inc. (Cat#K1603-1, Clontech, Palo Alto, CA, USA).

**Plasmids and cDNA libraries**

pAS2 and pACT2 series were originally constructed by Elledge lab (Durfee et al. 1993) and are available from Clontech laboratories (Cat#K1604-A, K1604-B). Other GAL4 based two-hybrid vectors, such as pGBT9 and pGAD424 series, were originally published by Field's lab and are available from both Stratagene, Inc. (Cat#235700,235722) and Clontech Laboratories, Inc. (Cat#K1605-A, K1605-B). LexA based two-hybrid vectors are available from Origene Technologies, Inc. (Cat#DPL-100, DPL-102). cDNA libraries for two-hybrid and one-hybrid screening are available from Origene, Stratagene, Clontech, and Invitrogen. Rigel also makes its own two-hybrid cDNA libraries from various tissues. All of these two-hybrid vectors share basic structures as shown in Fig. 4A.

The cDNA library needs to be amplified before screening. It is recommended that at least 200 15-cm plates should be used to grow up 10 million independent cDNA clones. High quality plasmid can be obtained with Qiagen DNA preparation kits.

Single-strand carrier DNA for yeast transformation is available from Origene or Clontech. Carrier DNA can also be made according to protocol by Ito et al. (1983).

**Special buffers**

Z buffer pH 7.0 (per liter)

| | |
|---|---|
| Na$_2$HPO$_4$•7H$_2$O | 16.1 g |
| NaH$_2$PO$_4$•H$_2$O | 5.50 g |
| MgSO$_4$•7H$_2$O | 0.246 g |
| KCl | 0.75 g |

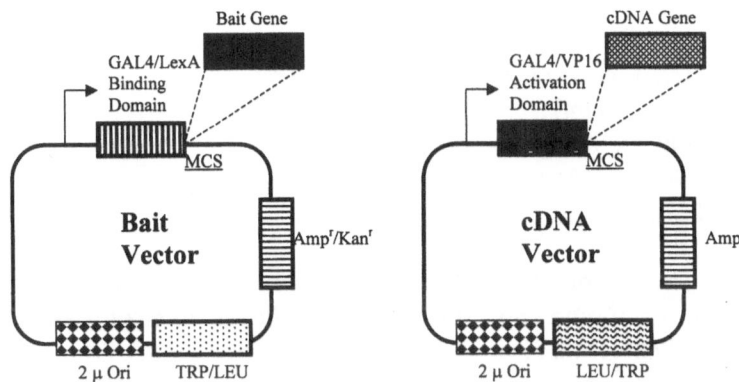

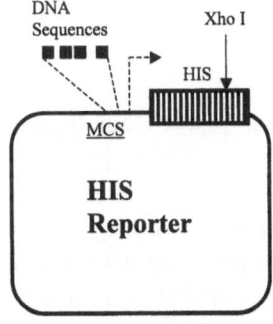

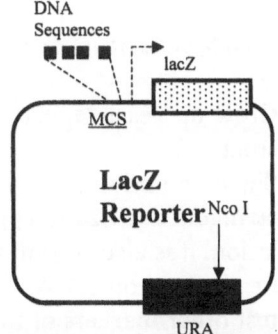

**Fig. 4.** Vectors used in the yeast two-hybrid and one-hybrid screening. **A** The two-hybrid vectors. Bait vectors can be pHybLex/Zeo (Invitrogen), pBD-GAL4 (Stratagene), pAS2–1 (Clontech), pGBT9 (Clontech), or pGilda (Origene, Clontech). *Arrows* indicate transcription of fusion proteins on either bait or cDNA vector. Binding domain can be either GAL4 or LexA. *MCS* represents multiple cloning sites, where either bait gene or cDNA fragments should be cloned. *2 μ Ori* represents yeast 2 μm replication origin. cDNA vectors can be pYESTrp2 (Invitrogen), pAD-GAL4 (Stratagene), pACT2 (Clontech), pGADGH (Clontech), pGAD424 (Clontech), or pJG4-5 (Origene). Activation domain can be GAL4, VP16, or other transcription activator. **B** The one-hybrid reporter vectors. DNA sequences of interest should be inserted into the multiple cloning sites (*MCS*) The enzyme used to linearize reporter vector for integration is shown by *solid arrow*. *Dashed arrow* indicates the transcription of either HIS or lacZ gene

Z buffer + X-Gal

| | |
|---|---|
| Z buffer | 1 ml |
| 20 mg/ml X-Gal | 40 µl |
| β-mercaptoethanol (optional) | 2 µl |

PEG/LiAc (10 ml)

| | |
|---|---|
| 50 % PEG(3350) | 8 ml |
| 10 × TE (pH 7.5) | 1 ml |
| 1 M LiAc | 1 ml |

**Equipment and others**  30 °C incubators and liquid nitrogen containers are required. Nylon membrane and Whatman filters for the lacZ color assay are available from Fisher Scientific. X-Gal is from either Promega (Cat#V3941, Madison, WI, USA) or Denville Scientific (Cat#CX-3000-3, Metuchen, NJ, USA). All plastic wares are from Fisher Scientific or VWR.

## Procedure

### Yeast two-hybrid system screening

1. Grow up yeast reporter strains on YPD plates from frozen stock:
   Since no antibiotics are added into the yeast medium, very stringent sterilization procedures are required during inoculation. It is also recommended that the yeast reporter strain be streaked on SD-W, SD-L, SD-H, SD-U, and SD-K plates to test other markers of the yeast before cDNA library screening. A reporter strain such as Y190 should be able to grow up on SD-K, SD-U, and SD-H plates, but not on SD-W, and SD-L plates. Growth on the SD-H plate is due to leaky expression of the HIS reporter gene.
   There are many reporter strains available from different resources. In general, Y190 consistently showed higher sensitivity than other yeast strains such as HF7 c. Yeast reporter strains with both lacZ reporter gene and HIS3 reporter gene are strongly recommended. HIS selection will ensure that only interaction-positive clones will grow, which makes colony picking much easier later.

2. Determine optimal 3AT concentration:
   3AT can be used to suppress background expression from the HIS reporter gene of Y190. 3AT concentration varies among different reporter strains and ranges from 0 mM (HF7 c) to 15 mM (Y190). To test the optimal concentration of 3AT, one yeast colony should be re-suspended in 10 ml of TE. 100 μl of the re-suspended yeast is spread on SD-H+0 mM 3AT, SD-H+5 mM 3AT, SD-H+10 mM 3AT, SD-H+15 mM 3AT, SD-H+25 mM 3AT, and SD-H+40 mM 3AT plates. Although 15 mM 3AT is sufficient to suppress background HIS expression of Y190, higher concentrations of 3AT (30–40 mM) are routinely used in our cDNA library screening.

3. Construct bait plasmid:
   pAS2/pACT2 series plasmids showed a higher level of sensitivity than pGAD424/pGBT9 series plasmids (Legrain et al. 1994; Estojak et al. 1995). The disadvantage of using pAS2 is the large size of this plasmid (8 kb), which may present a challenge to cloning large cDNA fragments into the plasmid. cDNA fragments should be fused to the C-terminal of the GaL4 binding domain in frame (Fig. 4A). The junction sequence between GAL4 and cDNA should have a GGG amino acid sequence to avoid any interruption of domain structure. Either full-length cDNA or partial fragments can used to generate bait plasmid.

4. Transform bait into yeast: 1st round:
   1 μg of bait plasmid is transformed into Y190 with the small-scale yeast transformation protocol (see Subprotocol section). Transformants should be plated on SD-W, SD-WH, and SD-WH+3AT(5–40 mM) plates. The lacZ color assay can also be done after colonies grow to a diameter of 1 mm. If colonies grow up on SD-WH+40 mM 3AT plates after 3 days of incubation and/or the lacZ color assay of these colonies show positive result after only 30-min incubation with X-Gal, the bait gene should be determined not suitable for two-hybrid screening without further modification. The bait gene itself may be able to activate transcription of reporter genes HIS/lacZ.
   Although co-transformation of bait plasmid and cDNA library can be done in a single step, co-transformation efficiency is at least tenfold lower than single plasmid transfor-

mation. The Mating approach may also be used to introduce the cDNA library into yeast cells containing the bait vector. Please refer to protocol published by Finley and Brent (1994).

5. Transform cDNA library: 2nd round:
   Y190 containing bait plasmid is grown up for the second round of transformation by cDNA library plasmid (see Subprotocol section). Incubation time after transformation varies significantly from 4 to 11 days.

6. Identify positive clones:
   Identification of positive clones needs experience. It should also be pointed out that background colonies at lightly populated area of the plate tend to grow bigger, occasionally reaching the size of a positive colony in a dense area on the same plate. The size of the positive colony should at least four times bigger than the neighboring background colonies. Positive colonies may also turn red faster.

7. Perform lacZ color assay:
   Positive colonies should be re-streaked to another SD-LWH+3AT plate to isolated single colonies for color assay and plasmid retrieval. See Subprotocol section for the lacZ color assay protocol. If a colony does not turn blue after a 4-h incubation, strong protein-protein interaction is highly unlikely. It is not recommended to pick positive clones after 12-h incubation, unless you know the protein-protein interaction you are studying is very weak.

8. Retrieve plasmids:
   There are several methods to retrieve plasmids from yeast, ranging from lyticase lysis to glass beads. The glass beads method is listed in the Subprotocol section. The electroporation method is by far the most efficient method to transform plasmids from yeast miniprep into *E. coli*. Bait and cDNA plasmid may carry different antibiotic selection markers to facilitate separation in *E. coli*. For example, Rigel's bait plasmid carries Kan$^r$ gene and the cDNA plasmid carries Amp$^r$ gene.

9. Verify positive clones:
   cDNA clones recovered from positive HIS/lacZ positive colonies should be re-transformed into yeast with another

non-specific bait control to rule out non-specific binding. In vitro protein binding assays and function assays should also be done to rule out false positive clones.

## Yeast one-hybrid system screening

1.  Construct HIS and lacZ reporter plasmids:
    Selection of a very-well-defined DNA sequence is the most important step for one-hybrid screening. Many DNA sequences lead to significant elevation of the basal expression levels of the reporter genes in yeast, even in the absence of the cDNA library. Multiple copies (~3) of the DNA sequences of interest should be inserted into the multiple cloning sites of both HIS reporter plasmid pHISi-1 and pLacZi (Clontech Cat#K1603-1; Fig. 4B).

2.  Grow yeast reporter strains on YPD plates from frozen stock:
    It is also recommended that the yeast reporter strain be streaked on SD-W, SD-L, SD-H, SD-U, and SD-K plates to test other markers of the yeast before cDNA library screening. A reporter strain such as YM4271 should be able to grow up on SD-K, SD-U, SD-H, SD-W, and SD-L plates.

3.  Integrate HIS reporter into yeast:
    To facilitate integration of pHISi reporter into yeast chromosome, pHISi-1 should be linearized at the Xho I site. Since pHISi-1 has no yeast replication origin and cannot survive in yeast without integration, no gel purification of digested plasmid is required. Transform 1 µg of digested plasmid into YM4271 using the small-scale yeast transformation protocol (see Subprotocol section). Use more plasmids if integration efficiency is low. Transformants should be plated on SD-H, and SD-H plates with different concentration of 3AT(5–40 mM) and incubated at 30 °C for at least 4 days.

4.  Determine optimal 3AT concentration:
    If more than 40 mM 3AT is needed to suppress transformants growth, the DNA sequences inserted into pHISi are not suitable for one-hybrid screening.

    **Note:** Integration efficiency of pHISi is very low. 20–100 colonies are expected on the SD-H plate.

5. Integrate lacZ reporter plasmid into yeast:
   Pick a colony from a SD-H plate from step 3 and freeze as single HIS reporter strain YM4271/H. Linearize pLacZi at the Nco I site. Transform 1 µg of linearized plasmid into yeast YM4271/H and plate transformants on SD-U plates. This step of integration is very efficient. Several hundred to a thousand colonies are expected to grow on each SD-U plates. Pick colonies and freeze as YM4271/HB.

6. Screen cDNA library for DNA binding protein:
   Transform 100–200 µg of cDNA library into YM4271/HB with large-scale yeast transformation protocol (see Subprotocol section). Transformants should be plated on SD-LH+3AT (concentration determined at step 4).

7. Identify positive clones:
   Same as step 6 in two-hybrid screening procedure.

8. Perform lacZ color assays:
   Same as step 7 in two-hybrid screening procedure.

9. Retrieve cDNA plasmid:
   Same as step 8 in two-hybrid screening procedure.

10. Verify positive clones:
    The DNA gel retardation assay and other function assays are required to verify one-hybrid screening results.

### Small scale yeast transformation (105 transformants/ µg DNA)

1. Inoculate one colony of yeast in 100 ml YPD (without plasmid) or corresponding selection medium (SD-W for Y190 with pAS2) at 240 rpm in a 30 °C shaker overnight.

2. Check $OD_{600}$ the next day. If $OD_{600}$ is between 0.6 and 1.0, the yeast can be used to prepare competent cells. Otherwise, dilute to $OD_{600}=0.4$ and grow another 3–4 h.

3. Spin down cells in two 50-ml plastic tubes at 3000 rpm at room temperature for 5 min. Remove medium.

4. Add 30 ml TE pH 7.5 and re-suspend the cell pellet on the vortex at high speed. Combine cell pellet.

5. Spin down cells again in at 3000 rpm at room temperature for 5 min.

6. Remove TE.

7. Estimate the size of the cell pellet and add TE up to a total volume of 0.9 ml. Re-suspend cells completely by pipetting up and down.

8. Add 100 µl 1 M LiAc and mix well by pipetting. Competent cells are ready.

   Note: Competent cells can be kept at room temperature for several hours without significant reduction of transformation efficiency, or at 4 °C overnight with a slight reduction of transformation efficiency.

9. In a clean Eppendorf tube, mix 1 µg of plasmid with 10 µl 10 mg/ml carrier DNA.

10. Add 100 µl competent cells from step 8 to the Eppendorf tube and mix well with the DNA.

11. Add 600 µl PEG/LiAc and mix well.

   Note: PEG/LiAc should be freshly made. Pre-mixed PEG/LiAc of up to 2 weeks old can also be used if transformation efficiency is not critical.

12. Incubate at 30 °C for 30 min with or without shaking.

13. Add 70 µl DMSO and mix well.

14. Incubate in 42 °C water bath for 15 min.

15. Put on ice for 2 min.

16. Spin down cells in an Eppendorf centrifuge at 8000 rpm for 1 min.

17. Remove supernatant.

18. Add 150 µl of TE to re-suspend cell pellet.

19. Plate on selection medium plate. (e.g. SD-W for Y190 transformed by pAS2).

20. Incubate in a 30 °C incubator for 2–3 days.

**Large scale cDNA library transformation (1–10 ¥ 10⁶ transformants/100 µg cDNA)**

1. Inoculate one colony of yeast in 200 ml YPD (one-hybrid screening) or corresponding selection medium (SD-W for two-hybrid screening) at 240 rpm in a 30 °C shaker overnight.

2. Check $OD_{600}$ the next day. If $OD_{600}$ is between 0.8 and 1.0, the yeast can be used to prepare competent cells. Otherwise, dilute to $OD_{600} = 0.6$ and grow another 3–4 h.

3. Spin down cells in a 250-ml bottles at 3000 rpm at room temperature for 5 min.

4. Remove medium.

5. Add 50 ml TE pH 7.5 and re-suspend the cell pellet on the vortex at high speed.

6. Spin down cells again in at 3000 rpm at room temperature for 5 min.

7. Remove TE.

8. Repeat steps 4–7 one more time.

Estimate the size of cell pellet and add TE up to a total volume of 1.8 ml. Re-suspend cell pellet completely by vortexing.

9. Add 200 µl 1 M LiAc and mix well by vortexing.

10. In a clean Eppendorf tube, mix 100–200 µg of plasmid with 200 µl 10 mg/ml carrier DNA.

11. Add DNA to competent cells drop by drop vortexing at 5000 rpm.

12. To ensure sufficient mixture, vortex at the highest speed for 30 s.

13. Add 12 ml PEG/LiAc and mix well.

Note: PEG/LiAc should be freshly made. Pre-mixed PEG/LiAc up to 2 weeks old can also be used if transformation efficiency is not critical.

14. Incubate at 30 °C for 30 min with shaking. Either an orbital shaker or rotator can be used.

15. Add 140 µl DMSO and mix well.

16. Incubate in 42 °C water bath for 15 min. Invert several times during incubation.

17. Put on ice for 5 min to chill.

18. Spin down cells at 3000 rpm in a bench-top centrifuge for 1 min.

19. Remove supernatant.

20. Add 20 ml TE and re-suspend cell pellet by vortexing.

21. Plate 400 µl on each 15-cm selection medium plates (50 plates total). SD-LWH+40 mM 3AT plates are used for Y190 strain two-hybrid screening; SD-LH+3AT plates are used for one-hybrid screening.

22. Plate 1 µl on a 10-cm plate of SD-LW for transformation efficiency control.

23. Incubate at 30 °C for up to 8 days until big colonies appear.


## LacZ color assay

1. Grow up fresh yeast colonies to a 1-mm diameter.

2. Fill a container (e.g., ice bucket) with liquid nitrogen.

3. Use a nylon membrane to lift colonies up from the plate. No special replica-plating device is needed. Simply press the nylon membrane to the plate.

4. Immerse the nylon membrane (Cat# N04HY08250, N04HY13250, Fisher Scientific, PA, USA) colony-side face down into the liquid nitrogen.

5. Wait for 20 s, remove the nylon membrane and allow to dry on a paper towel for 5 min.

6. In a 10-ml tube, add 40 µl X-Gal to each ml of Z buffer.

7. Add 1.5 ml Z buffer/X-Gal solution to a clean 10-cm petri dish. For a 15-cm diameter petri dish, add 4 ml Z buffer/X-Gal solution.

8. Add a Whatman circle to the petri-dish, ensuring it is evenly soaked and any air bubbles are squeezed out.

9. Use forceps to transfer the dried nylon membrane with colony-side facing up to lie over the soaked Whatman circle (Cat#09-805C, Fisher Scientific, PA, USA). Make sure there are no air bubbles in between the membrane and the circle.

10. Incubate the petri-dish with lid on in 37 °C until blue color is visible.

## Yeast plasmid mini-isolation

1. Inoculate 3 ml of selection medium ( e.g., SD-L for cDNA library plasmid pACT ) with a yeast colony.

2. Incubate in a 30 °C shaker or rotator overnight or until confluent.

3. Spin down yeast in a bench-top centrifuge at 3000 rpm at room temperature.

4. Remove medium and re-suspend pellet in 200 μl lysis buffer. Transfer to an Eppendorf tube.

5. Add 200 μl volume glass beads.

Note: The lid of the Eppendorf tube can be used as a scoop to collect 200 μl glass beads.

6. Add 200 μl phenol/chloroform/isoamyl alcohol (25:24:1).

7. Vortex at the highest speed for 3 min.

8. Spin in micro-centrifuge at 14,000 rpm for 10 min.

9. Transfer top aqueous layer to another Eppendorf tube, add 20 μl 3 M NaAc and 500 μl ethanol. Precipitate should be visible immediately.

10. Put the Eppendorf tube into a dry ice bath for 15 min or until frozen.

11. Spin in a micro-centrifuge at 14,000 rpm for 10 min.

12. Remove supernatant and dry pellet.

13. Wash pellet with 100 µl of 80 % ethanol, and dry the pellet in air.

14. Re-suspend pellet in 30 µl $H_2O$ and use 1 ml for electroporation to transform *E. coli*.

## Results

**Two-hybrid screening:** p21 peptide was cloned into pAS2-1 to screen for proteins that can bind to p21 peptide. Results are listed below.

| cDNA library | Human lymphocyte |
| --- | --- |
| Bait vector | pAS2-1 |
| Bait protein | p21 peptide |
| Yeast strain | Y190 |
| Number of transformants | 15 million |
| HIS+/lacZ+ clones | 2 |
| Clone identity | PCNA (2) |

PCNA is a published binding protein of p21 peptide. Yeast one-hybrid screening results were previously published (Luo et al. 1996).

*Acknowledgements.* We would like to thank Mary Shen, who constructed many two-hybrid cDNA libraries in Rigel for our screening effort, and Dr. Mengsheng Qiu, who provided useful suggestions for improvement of the two-hybrid screening. We would also like to thank Karla Blonsky for providing editorial help.

## References

Chevray PM, Nathans D (1992) Protein interaction cloning in yeast: Identification of mammalian proteins that react with the leucine zipper of Jun. Proc Natl Acad Sci USA 89:5789–5793
Chien CT, Bartel PL, Sternglanz R, Fields S (1991) The two-hybrid system: a method to identify and clone genes for proteins that interact with a protein of interest. Proc Natl Acad Sci USA 88:9578–9582

Durfee T, Becherer K, Chen PL, Yeh SH, Yang Y, Kilburn AE, Lee WH, Elledge S (1993) The retinoblastoma protein associates with the protein phosphatase type 1 catalytic subunit. Genes Dev 7:555–569

Estojak J, Brent R, Golemis EA (1995) Correlation of two-hybrid affinity data with in vitro measurements. Mol Cell Biol 15:5820–5829

Fields S, Song O (1989) A novel genetic system to detect protein-protein interactions. Nature 340:245–247

Finley R, Brent R (1994) Interaction mating reveals binary and ternary connections between Drosophila cell cycle regulators. Proc Natl Acad Sci USA 91:12980–12984

Gstaiger M, Knoepfel L, Georgiev O, Schaffner W, Hovens, CM (1995) A B-cell coactivator of octamer-binding transcription factors. Nature 373:360–362

Ito H, Fukada Y, Murata K, Kimura A (1983) Transformation of intact yeast cells treated with alkali cations. J Bacteriol 153:163–168

Legrain P, Dokhelar MC, Transy C (1994) Detection of protein-protein interactions using different vectors in the two-hybrid system. Nucleic Acids Res 22:3241–3242

Lehming N, Thanos D, Brickman JM, Ma J, Maniatis T, Ptashne M (1994) An HMG-like protein that can switch a transcriptional activator to a repressor. Nature 371:175–179

Li JJ, Herskowitz I (1993) Isolation of ORC6, a component of the yeast origin of recognition complex by a one-hybrid system. Science 262:1870–1873

Luo Y, Stile J, Zhu L (1996) Cloning and analysis of DNA-binding proteins by yeast one-hybrid system and yeast one-two-hybrid system. BioTechniques 20:564–568

Mendelsohn AR, Brent R (1994) Applications of interaction traps/two-hybrid systems to biotechnology research. Curr Opin Biotechnol 5:482–486

Shang J, Luo Y, Clayton D (1997) *Backfoot* is a novel homeobox gene expressed in the mesenchyme of developing hind limb. Dev Dyn 209:242–253

Strubin M., Newell JW, Matthias P (1995) OBF-1, a novel B cell-specific coactivator that stimulates immunoglobin promoter activity through association with octamer-binding proteins. Cell 80:497–506

Wang MM, Reed RR (1993) Molecular cloning of the olfactory neuronal transcription factor Olf-1 by genetic selection in yeast. Cell 74:205–214

Wilson TE, Fahrner TJ, Johnston M, Milbrandt J (1991) Identification of the DNA binding site for NGFI-B by genetic selection in yeast. Science 252:1296–1300

Zervos A, Gyuris J, Brent R (1993) Mix1, a protein that specifically interacts with Max to bind to Myc-Max recognition sites. Cell 72:223–232

## Suppliers

Denville Scientific, Inc., P.O. Box 4588, Metuchen, New Jersey 08840, USA (Tel.: +1-908-7577577, Fax: +1-908-7577551)

Fisher Scientific, 771 Forbes Avenue, Pittsburgh, Pennsylvania 15219–4785, USA (Tel.: +1-800-7667000, Fax: +1-800-9261166)

Stratagene, 11011 North Torrey Pines Road, La Jolla, California 92037, USA (Tel.: +1-800-4245444)

Invitrogen, 1600 Faraday Avenue, Carlsbad, California 92008, USA (Tel.: +1-800-9556288, Fax: +1-619-6037201)

Clontech Laboratories, 1020 East Meadow Circle, Palo Alto, California 94303–4230, USA (Tel.: +1-800-6622566, Fax: +1-800-4241350)

Promega, 2800 Woods Hollow Road, Madison, Wisconsin 53711–5399, USA (Tel.: +1-800-3569526, Fax: +1-800-3561970)

Qiagen, 28159 Avenue Stanford, Santa Clara, California 91355, USA (Tel.: +1-800-4268157, Fax: +1-800-7182056)

OriGene Technologies, 13 Taft Court, Suite 111, Rockville, Maryland 20850, USA (Tel.: +1-888-2674436, Fax: +1-301-3409254)

Sigma Chemical Company, P.O. Box 14508, St. Louis, Missouri 63178–9916, USA (Tel.: +1-800-3253010, Fax: +1-800-3255052)

# High-Throughput Library Screening by Fluorescent Hybridizations on Gridded Membranes

HUGUES ROEST CROLLIUS, JOHN O'BRIEN, HANS LEHRACH

## Introduction

Large-scale structural or functional genome analysis often deals with entire genomic or cDNA clone libraries that must be screened repeatedly with hundreds to thousands of DNA probes. High throughput protocols have been established to achieve this, making use either of massively parallel PCR screens or of hybridization experiments on high density gridded libraries. The widespread use of the latter approach has been hampered by the need to manipulate high doses of radioactivity, and to prepare good quality high density filters. We describe here a set of protocols that use "cold" fluorescent signal detection on high density filters that allows us to routinely screen about 50 probes per day on membranes each carrying 55,000 individual clones. This fast and cost-effective alternative to massive parallel PCR protocols requires little specialized equipment and can be operational within a few days.

H. Roest Crollius (e-mail: hrc@genoscope.cns.fr, Tel: +33-(0)1-60872564, Fax: +33-(0)1-60872589)
Genoscope, CNRS UMR8030, 2, rue Gaston Crémieux, 91057 Evry Cedex, France
John O'Brien
Max-Planck-Institut für Molekulare Genetik, Ihnestrasse 73, 14195 Berlin, Germany
H. Lehrach
Max-Planck-Institut für Molekulare Genetik, Ihnestrasse 73, 14195 Berlin, Germany

Springer Lab Manual
R.C. Bird, B.F. Smith (Eds.) Genetic
Library Construction and Screening
© Springer-Verlag Berlin Heidelberg 2002

The protocols described here can be used in a variety of probe/library combinations and have been used successfully to hybridize single copy or repeated probes on cDNA and genomic libraries (plasmid; cosmid; PAC, P1-derived artificial chromosome; BAC, bacterial artificial chromosome). The first protocol describes how to fix DNA from *E. coli* colonies that have been gridded at high density on nylon membranes. It assumes that one has the use of a robotic gridding machine to actually create the array of colonies on the membranes. Alternatively, it is possible to find ready-to-use library membranes from academic or commercial sources (see Protocol 1, comments), although any large-scale project would need to generate an ample provision of self-made membranes to remain cost-effective. The second protocol is a simple PCR reaction that yields a labeled probe by incorporation of dUTP attached to a digoxigenin (DIG) molecule during the polymerization. The digoxigenin/anti-digoxigenin system to label and detect nucleic acids (Höltke et al. 1995) is an alternative to biotin/(strept)avidin and we found that it produces the lowest background and most reproducible results. Alternative methods to incorporate DIG include random hexamer priming with Klenow polymerization or nick translation, but PCR is the easiest to carry out on a large scale and includes a convenient control for the incorporation of label that is not applicable with the other labeling techniques . The third protocol is the hybridization itself and is geared towards a large-scale application. Obviously, several steps can be adapted if experiments are carried out on a small scale, and these include the use of an air gun to apply the various solutions which can be replaced by traditional hybridization bottles (see Fig. 5). The principle of the detection method is based on the binding of an anti-DIG antibody to the labeled probe, which is bound to an alkaline phosphatase moiety. The dephosphorylation of a substrate molecule (Attophos) generates a fluorescent compound (Attofluor) that absorbs light at 430 nm and emits at 560 nm (Fig. 1). Attophos is the benzothiazole-derivative 2'-(2-benzothiazolyl)-6'-hydroxybenzothiazole phosphate and its use as a substrate for alkaline phosphatase has already been demonstrated in a number of applications including solution assays (Cano et al. 1992) and hybridization screens on high density membranes (Maier et al. 1994; Scholler et al. 1998).

The unusually large Stokes shift of 140 nm is a key element of the system. It considerably reduces the level of background

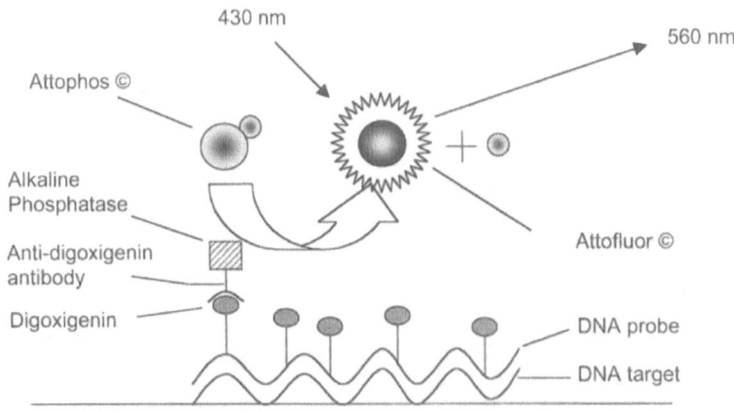

**Fig. 1.** Schematic representation of the detection system described in this chapter. After the DIG-labeled probe has been specifically bound to its target by hybridization, an anti-DIG antibody conjugated to an alkaline phosphatase molecule is applied, followed by the Attophos substrate. After a few minutes to several hours, a fluorogenic molecule is generated by dephosphorylation, which emits light at 560 nm when excited by 430 nm light (optimally) or long-wave UV light (in practice)

which is due to fluorescent emission of the substrate that has not been dephosphorylated. Although the optimal excitation maximum is in the visible range, we have successfully used long wavelength (365 nm) UV lamps instead of complicated optics that only emit a narrow bandwidth such as laser beams. In fact, the same equipment that is commonly used for ethidium bromide-stained agarose gels can be used without modifications to detect Attofluor signals. Large-scale applications might, however, require a more optimized setup, as described in Protocol 3. This includes a high resolution Charge Coupled Device (CCD) camera to facilitate image analysis, and a dedicated light source and light-tight enclosure.

## ░ Outline

Figure 2.

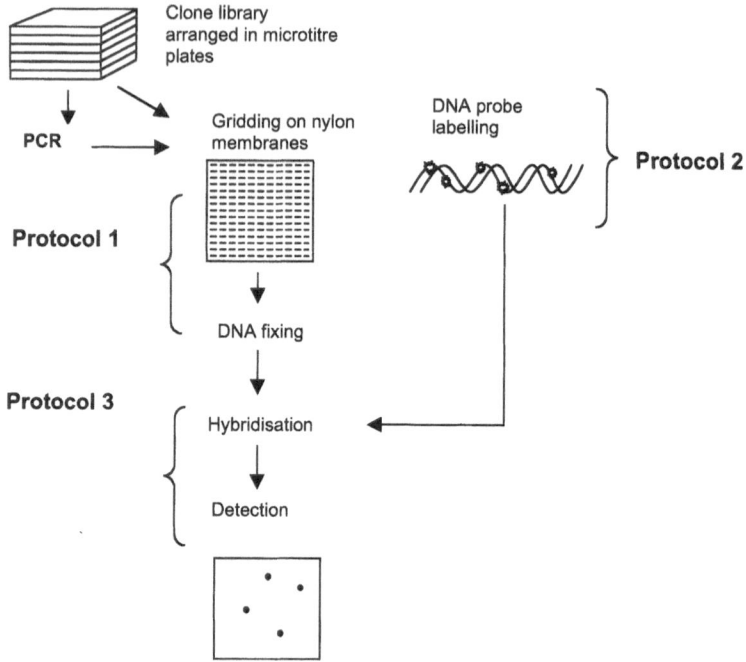

**Fig. 2.** A clone library stored in microtiter plates is thawed and gridded onto nylon membranes. After fixing the DNA (Protocol 1), a probe is prepared (Protocol 2) and hybridized (Protocol 3) onto the membrane.

## ░ Materials

**Equipment** – A heating water bath with temperature range up to 95 °C and bath surface at least 25 × 25 cm
– Robotic gridding machine ( e.g., QBOT, Genetix)
– Air compressor, 8 bar, 9 l tank is sufficient.
– Air gun, bought from local DIY shop, used for painting large surfaces.
– Plastic laminator (e.g., GMP MR-12 desktop pouch laminator, GMP Ltd.)
– Roll for plastic lamination

- CCD camera (e.g. SPOT RT, Diagnostic Instruments Inc.)
- Plastic consumables (square trays, plastic pin tool, microtiter plates, Genetix Ltd.)

**2YT media**

**Buffers and solutions**

16.0 g/l  bacto-tryptone
10 g/l    bacto-yeast extract
5 g/l     NaCl
(15 g/l   bacto-agar)
pH to 7.0, autoclaved.

Add the appropriate antibiotic before use.

**Denaturing solution**
0.5 M    NaOH
1.5 M    NaCl

**Neutralizing solution**
1 M      Tris-Cl pH 7.0
1.5 M    NaCl

**Pronase solution**
50 mM    EDTA
100 mM   NaCl
50 mM    Tris-Cl pH7.4
1 %      Sarcosyl

**10 × PCR buffer**
500 mM   KCl
100 mM   Tris-Cl pH 8.5

Autoclave before use

**10 × dNTP mix for probe labeling**
2 mM     dATP
2 mM     dCTP
2 mM     dGTP
1.9 mM   dTTP
0.1 mM dUTP-DIG

**1 × TAE buffer**
40 mM  Tris-acetate pH 8.0
1 mM    EDTA

**1 × TE buffer**
10 mM  Tris-Cl pH 7.4
1 mM    EDTA

**Hybridization buffer**
500 mM  Na Phosphate pH 7.2
5 %  SDS

**Stringency wash buffer**
40 mM  Sodium Phosphate
0.1 %  SDS

**Blocking buffer**
40 mM  Na Phosphate pH 7.2
150 mM  NaCl
5 %  Milk powder (0 % fat)

**Antibody wash buffer**
40 mM  Na Phosphate pH 7.2
150 mM  NaCl

**Substrate buffer**
2.4 M  Diethanolamine
0.06 mM  $MgCl_2$
0.6 mg/ml  Attophos

**Stripping solution 1**
400 mM  NaOH
0.1 %    SDS

**Stripping solution 2**
100 mM  NaPi
0.1 %    SDS
0.1 M    Tris-Cl pH 7.4

## Subprotocol 1
## Preparation of gridded colony filters

■ ■ Procedure

1. Thaw the microtiter plates by laying them on a bench at room temperature. Wipe any condensation which may form on the lids with an autoclaved piece of blotting paper.   **Layout of the membrane**

2. Cut a piece of positively charged nylon membrane to the desired size and write a suitable identification on the side with a pencil.

**Note:** (1) The standard format for use with robotic equipment is 22.2 × 22.2 cm. Pre-cut membranes can be purchased from most suppliers. This format maximizes the use of the surface of the membrane. (2) While handling membranes, wear non-powdered gloves.

3. On a square plastic support, lay in succession, two pieces of blotting paper slightly oversized compared to the membrane, soaked with 2YT media (with appropriate antibiotic). Expel any air bubbles and drain the excess media by rolling a test tube on the paper. The paper must be wet but no media should be dripping from it when tilting the support at an angle.

4. Lift the membrane with blunt-end forceps by two diagonally opposed corners, soak it in 2YT media (with appropriate antibiotic) and position it carefully on the center of the wet blotting paper. As previously, expel air bubbles and drain excess media (Fig. 3)

5. If using a robotic device for the transfer of the bacterial inoculum from the microtiter plates to the membrane, follow the manufacturer's instructions. If transferring by hand, use a sterile plastic or metallic tool with 384 or 96 pins. Dip the tool in the wells of the microtiter plates, and apply once carefully on the membrane.

6. Once the bacterial clones have been gridded, lift the membrane with forceps and transfer it onto a square plastic bioas-

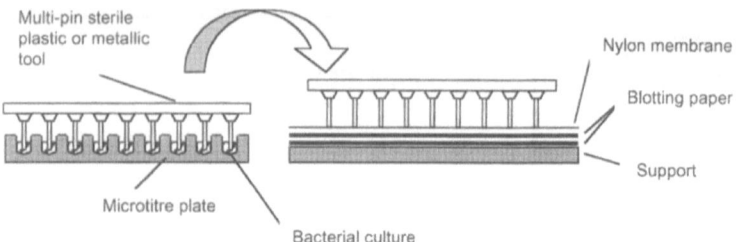

**Fig. 3.** Schematic representation of a colony transfer between a microtiter plate (*left*) to a nylon membrane that has been laid out on blotting paper soaked in growth media

say tray containing 2YT agar (with appropriate antibiotic). The surface of the agar must be slightly dry. To this end, place the bioassay trays containing the set agar open in a sterile laminar flow hood for 30 min prior to use.

7. Incubate the membranes at 37 °C, membrane facing down, for 12–16 h.

**Fixing of the DNA**

8. In the lid of a bioassay tray, place a piece a blotting paper slightly oversized compared to the membrane, soaked with denaturing solution. Drain the excess solution.

9. Holding the membrane by diagonally opposed corners with forceps, transfer the membrane from the agar plate to the blotting paper. Ensure no air bubbles are trapped under the membrane. Leave 4 min.

10. Transfer the membrane to a fresh piece of blotting paper soaked in denaturing solution. Holding both the membrane and the paper together, transfer onto a glass plate that has been placed in a boiling water bath, above the surface of the water. Ensuring the inside of the lid is dry, close the water bath and leave 4 min (Maas 1983).

11. Remove the lid carefully without dropping any water that may have condensed on the inside surface, and place the membrane on a fresh piece of blotting paper soaked with neutralizing solution. Leave 4 min.

12. Carefully submerge the membrane in an excess volume of pronase buffer containing 0.25 mg/ml of pronase that has been pre-warmed to 37 °C. Incubate 45 min at 37 °C.

13. Place the membranes DNA side-up on a piece of blotting paper, and leave at least 24 h to dry.

14. Cross-link the DNA to the nylon by a 120 mJ UV treatment. Afterwards, store in a dry place at room temperature sandwiched between two fresh pieces of blotting paper.

15. Stick the membranes onto a 350-μm-thick plastic backing using a commercial pouch laminating machine (Bancroft et al. 1997).

## ▉ ▉ Troubleshooting

- The growth of colonies on the filter is uneven, with defined areas empty: air bubbles were trapped between the agar and the nylon membranes, preventing access of the bacteria to the media. Ensure no air bubbles are trapped when placing the membrane on the agar. This can be avoided by ensuring that the agar surface is dry.

- The growth of colonies on the filter is uneven, with a general patchy aspect: the microtiter plates containing the clones may have been too old and many clones were unable to grow, or the library itself contained many empty wells. Make a fresh copy of the library, or rearrange the clones within the plates to ensure that all wells contain a bacterial culture.

- Colonies do not form a regular grid but have merged, or form a bacterial bed: the surface of the agar was not dry enough. After the agar has solidified, bio-assay trays should be placed with their lid open in a sterile hood over a constant air flow for at least 30 min.

## ▉ ▉ Comments

- Some applications may require that DNA be spotted instead of bacterial colonies, for instance when a high amount of target DNA is needed to obtain a signal or when the bacterial genome may interfere with the hybridization. In such cases, the DNA is most easily obtained by PCR amplification of each insert individually in microtiter plate format, provided the insert is small enough. DNA samples are gridded as described, but on dry membranes taped on a rigid support, and the only treatment required is a short (2 min) denaturation step on blotting paper soaked in denaturing solution, followed by a neutralization, and UV cross-linking as described in step 14.

- The processing of the membrane described here may seem over complicated compared to protocols supplied by membrane manufacturers. Large scale projects, however, require gridded colony membranes that give the best signal-to-noise ratio possible and can be used over and over without loss of signal. The protocol given here achieves this result, with membranes that can be re-hybridized 10 to 15 times with the accompanying hybridization protocol.

- Gridded colony membranes from a wide range of genomic and cDNA libraries can be obtained from the German Human Genome Project Resource Centre in Berlin (http://www.rzpd.de/). Gridded membranes can also be obtained from commercial sources (e.g., Genome Systems Inc. or Research Genetics) and other resource centers (e.g., HGMP, UK). These membranes can be used directly in the hybridization protocol described here.

## Subprotocol 2
## Labeling of probe

### ■ ■ Procedure

1. For a typical PCR amplification in 50 µl, prepare a reaction mix as follows:
   1 × Reaction buffer
   1 × Mix dNTP
   1.5 mM    $MgCl_2$
   1 unit     Taq polymerase
   200 nM    Primer1
   200 nM    Primer2
   50 ng     DNA template

2.  After distributing the mix and DNA templates, place the tubes in a thermocycler (e.g., PTC 200, MJ Research Inc.) and execute the program that is recommended for the primers and DNA template. Typically, this would be as follows:
    –   94 °C for 2 min
    –   30 cycles at 94 °C for 30 s, 55 °C for 1 min, 72 °C for 1 min
    –   72 °C for 3 min

3.  To check the amplification, separate 5 µl of the amplification products by electrophoresis on a 1.5 % agarose gel in 1 × TAE buffer. **Note:** It is advisable to run a labeled and unlabeled sample side by side, as a slight increase in size of the labeled fragment will confirm that it has incorporated DIG-dUTP (Fig. 4).

4.  The labeled sample may be kept at –20 °C for up to 6 months.

## ▦ ▦  Results

Figure 4.

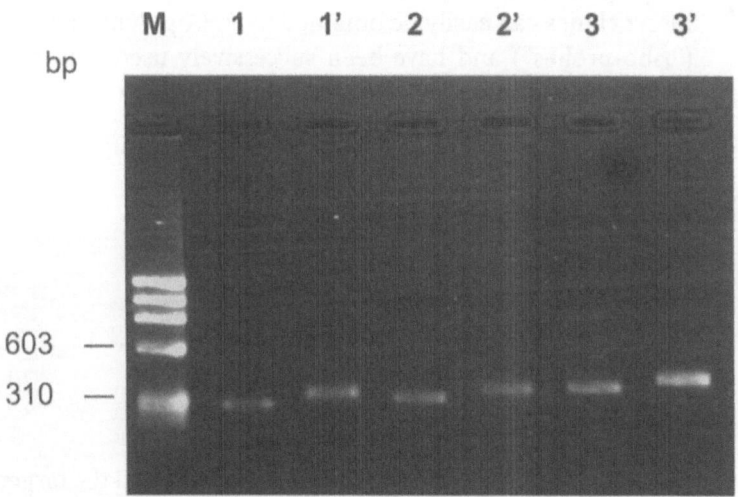

**Fig. 4.** Agarose gel containing *Tetraodon nigroviridis* STS amplifications by PCR with and without incorporation of digoxigenin. *M* ΦX174/HaeIII marker. Samples *1*, *2*, and *3* PCR products without DIG. Samples *1'*, *2'*, and *3'* the same fragments with 1:10 ratio of DIG-dUTP:dTTP in the PCR mix. A marked increase in fragment size is noticeable in DIG-labeled products

### ▦ ▦ Troubleshooting

- The PCR reaction does not yield an amplification product while the same PCR reaction without DIG-dUTP does: there may be too high a concentration of DIG-dUTP in the nucleotide mix which would partially inhibit the polymerase. Ensure that the ratio of DIG-dUTP:dTTP is below 35:65. The 10:90 ratio given in the protocol is sufficient for most applications while ensuring an efficient amplification.

- The size shift is not apparent between the labeled and unlabelled product: the shift is slight and only visible if the samples are run side by side. Absence of a size shift is not necessarily a sign that the probe is not labeled, and may depend on the resolution of the agarose gel.

### ▦ ▦ Comments

- The advantage of hybridization probes over PCR-based assays, is that in many cases it is not necessary to determine the sequence of the probe first. End sequences from large insert clones can easily be obtained by RNA polymerization ("ribo-probes") and have been successively used with this protocol. Whole inserts from cDNA clones can easily be obtained by PCR using vector primers.

---

### Subprotocol 3
### Hybridization and detection

### ▦ ▦ Procedure

**Pre-annealing of the probe**

1. This step is optional and depends on the probe and the target clone library, and is therefore highly application-dependent. This step is not necessary for single copy probes hybridized to genomic or cDNA libraries. It is, however, necessary for a more complex probe, such as a human whole cosmid which would contain repetitive elements. Mix in a tube 200 ng of

cosmid DNA, 20 μg of total human DNA, and 120 mM Na Phosphate pH 7.2, make up to 100 μl with 1 × TE. Pierce the cap of the tube with a needle or use a screw cap tube. Place in a boiling water bath for 5 min, centrifuge briefly to collect the condensation and incubate at 65 °C for 3 h.

2.  This step is also optional and depends on the application. Again, it is not required for single copy probes on genomic or cDNA filters. If it must be performed because repetitive sequences must be blocked, this step is best performed in a glass bottle or a sealed plastic bag and in fact prevents the use of the air gun as described below for the hybridization. As far as possible, it is recommended to replace this pre-competition step by the pre-annealing of the probe in step 1.

    **Pre-competition of the filter**

3.  Wet the membrane by submerging it shortly in hybridization buffer, and drain the excess buffer on blotting paper. Place the filter in an extraction hood (as used for solvents).

    **Hybridization**

4.  If the probe has been pre-annealed in step 1, it is ready to use. If not, denature the probe (about 1 μg for a 22.2 × 22.2 cm filter) in 100 μl of 1 × TE in a boiling water bath for 5 min. Alternatively if a large number of samples are to be hybridized, a thermocycler may be used to bring the samples to 95 °C for 8 min. **Note:** the use of NaOH to denature the probe is not possible since the digoxigenin molecule is attached to the dUTP by an alkali-labile bond.

5.  Directly add the denatured probe to 3 ml of pre-warmed hybridization buffer at 65 °C. Load the air gun with this solution and spray evenly over the membrane (Fig. 5). Lay a piece of cling film over the membrane unless laminated membranes are used, in which case membranes can simply be piled up on top of each other. Repeat step 5 for each probe.

6.  Wrap the pile of hybridized membranes in cling film and place in a sealed plastic bag to avoid evaporation. Incubate 6–16 h at 65 °C.

7.  Place all the membranes in a plastic box and wash twice for 10 min with agitation at 65 °C in an excess volume of stringency wash buffer pre-warmed to 65 °C. Approximately 100 ml per 22.2 × 22.2 cm membrane is recommended.

    **Washing of filters**

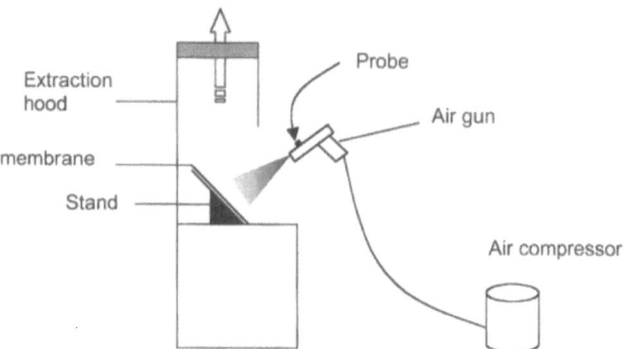

**Fig. 5.** Setup used for hybridization. The membranes are placed on a tilted stand and sprayed with the buffer containing the denatured probe. This setup is used for all subsequent treatments of the filter (antibody and substrate).

**Generation of fluorescent signals**

8. Replace the last stringency wash buffer with the same volume of blocking buffer, and leave 5 min at room temperature with gentle agitation.

9. Drain each membrane on blotting paper and place them successively on a piece of cling film in an extraction hood. As in step 5, evenly spray 3 ml of blocking buffer containing 0.1 units/ml of anti-DIG antibody conjugated to alkaline phosphatase.

10. Wrap in cling film and leave at room temperature for 45 min.

11. Place all the membranes in a container and wash three times 5 min in an excess volume of antibody wash buffer, at room temperature and with gentle agitation.

12. Replace the last wash with 50 ml per filter of 0.1 M diethanolamine pH 9.5, 1 mM $MgCl_2$. Leave without agitation for at least 2 min until ready to proceed.

13. Drain each membrane on blotting paper and place them successively in an extraction hood. With the air gun, evenly spray 3 ml of Attophos substrate. Wrap the stack of membranes in cling film and incubate 1–12 h at 37 °C.

**Signal detection**

14. Place the membrane under a CCD camera in a dark cubicle equipped with two long wave (365 nm) UV lamps (8 W each) as described in Fig. 6. The CCD camera should be equipped with a 500 nm cut-off filter (identical to filters used for ethid-

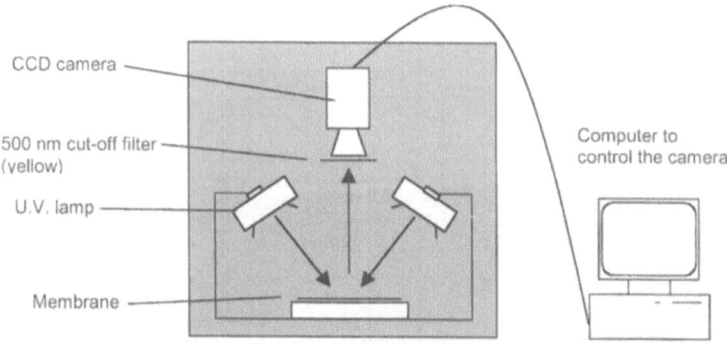

**Fig. 6.** Schematic representation of the setup required for optimal signal detection. Two 8 W long wave UV lamps are mounted at a 45° angle on each side of a flat platform on which the membrane will be deposited. The distance between the lamps and the membrane is about 30 cm. The CCD camera is placed directly above the membrane. The complete setup is placed in a light-tight enclosure

ium bromide-stained agarose gels). Use the camera control functions to acquire the best image of the fluorescent signals (Fig. 6). **Note:** under long-wave UV, the signals are clearly visible to the naked eye. **Caution:** A protective face shield should be used when working with UV light.

15. After the signal acquisition, the membrane can be stored at –20 °C to keep the signal intact several days. Otherwise the probe and fluorescent signals should be removed in preparation for the next hybridization.

16. Submerge the membrane a few seconds in stripping solution 1. **Note:** The Attofluor molecule will turn bright yellow under these alkaline conditions. Drain the membranes on blotting paper.

17. Wash twice for 5 min each in an excess volume of stripping solution 2.

18. Wash for at least 1 h in hybridization buffer at 65 °C. **Note:** the purpose of this step is to remove the fluorescent signal itself, while the probe and the alkali-labile digoxigenin have already been removed in step 16.

19. Drain the membrane on blotting paper and store at room temperature until the next hybridization, or re-hybridize directly.

## ◼ ◼ Results

Figure 7.

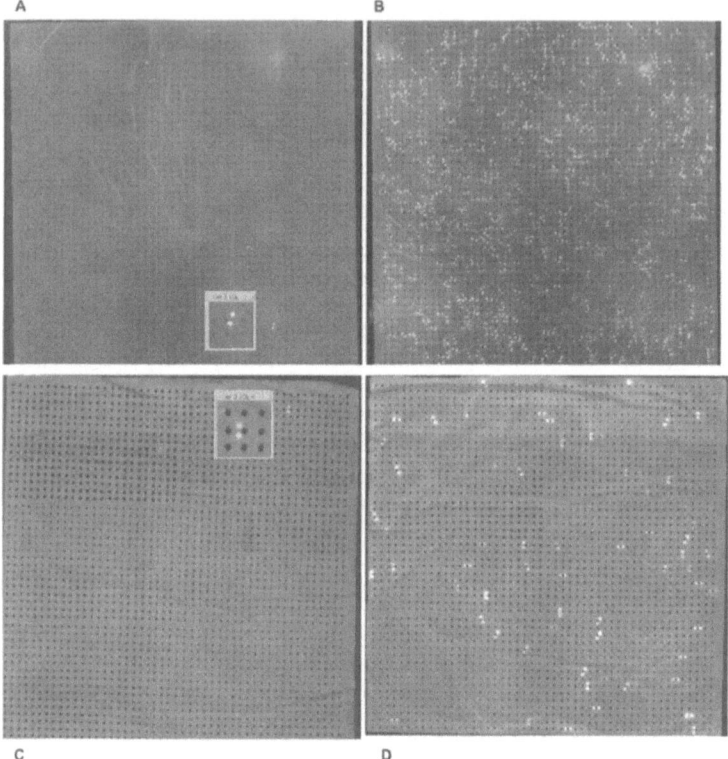

**Fig. 7.** Examples of hybridization experiments that illustrate the variety of probe/library combinations that can be analyzed with the protocols described here. **A** A single copy cDNA probe was hybridized to a gridded cDNA library (55,000 colonies on the figure) enriched for human X chromosome sequences (B. Korn, unpubl.). The *inset* shows a magnified region containing a duplicate positive colony, and clearly demonstrates the total absence of background and the excellent signal-to-noise ratio. **B** The same library as in A was hybridized with a human ALU probe to reveal those clones that contain repetitive sequences and which should be excluded from further analysis. **C** A *Tetraodon nigroviridis* (freshwater puffer fish) BAC library (H. Roest Crollius et al. 2000) was hybridized with a single copy STS probe generated by PCR from a BAC end sequence. **D** The same library as in C hybridized with a ribosomal 28S probe. This membrane contains 18,000 BAC clones and the experiment revealed 64 positive duplicate colonies.

## ▪ ▪ Troubleshooting

- There is no signal after leaving the membranes overnight at 37 °C. It is unlikely that the signal will develop after this time. Several factors may have caused the absence of signals.

  (1) absence of target for the probe in the library. Increase the coverage of the library or switch to a different library cloned in a different vector, or constructed with an alternative cloning strategy.

  (2) Probe concentration too low. The fluorescent hybridization protocol described here requires a higher probe concentration compared to a radioactive hybridization with the same probe/membrane combination. It is not unreasonable to increase the probe concentration to 1 µg/ml of hybridization buffer to obtain a signal.

  (3) Hybridization temperature too high. The 65 °C hybridization temperature indicated here works well for probes 200 bp and higher, while ensuring a specific binding of the probe. Shorter probes may need to be hybridized at lower temperatures. Try decreasing the temperature by 5 °C increments.

  (4) Too stringent post-hybridization washes. The temperature of the washes should be the same as the hybridization temperature. To reduce their stringency, shorten the incubation times in step 7, and/or replace the first wash by a rinse in the same buffer but at room temperature.

  (5) The pH of the diethanolamine buffer in step 12 is too high. This may cause the digoxigenin molecules to be released from their dUTP moiety. Check that the pH of this buffer is in the pH 9.5–10.0 range.

- There is a high colony background together with the expected signal. The advantage of fluorescent detection systems is that the increase in signal strength due to alkaline phosphatase turnover can be monitored in real time simply by checking the membrane under UV light. In principle, the expected signal should develop first, followed eventually by a colony background. It is possible to limit the contribution of the background by acquiring the image earlier. Alternatively, the background may be due to:

(1) A too high probe concentration. Repeat the experiment but reduce the probe concentration

(2) A too mild post-hybridization wash. Repeat the experiment but increase the length of the two washes in step 7.

(3) A contaminating sequence in the probe that binds to either the cloning vector or the host genome. Repeat the experiment with a purified probe or envisage a pre-annealing step to block the contaminating sequence.

- There are defined areas of high background on the membrane.

  (1) Powdered gloves worn while manipulating the membranes, even during the processing protocol, will leave permanent marks after detection.

  (2) The anti-DIG antibody and/or the substrate may have been unevenly applied with the air gun. This can be the case if the gun was placed too close to the membrane when spraying. It is worth spraying diluted ink on white paper to visualize and calibrate the distribution of the liquid by the air gun before using it in hybridization experiments.

- There are bright fluorescent spots on the membrane that are not due to the expected signal. Non-dissolved particles of powdered milk may create this effect. Ensure that the milk is completely dissolved in the blocking buffer. If particles remain, try centrifuging the blocking buffer for a few minutes at 2000 $g$, and use the supernatant, or switch to another brand of milk.

## Comments

- Image analysis. The final result of a hybridization experiment is an image stored on a computer, containing a few bright "positive" spots among thousands of "negative" spots. While it is possible to work out the position of the positive clones on the grid manually, and thus find the address of the corresponding clone in the original library plate, this work can be partially automated. This is especially important for large scale projects where several hundred images must be scored. Purpose-built commercial software exists that overlays a grid on the image, and either expects the operator to click on the positives with the computer mouse, or will work out automatically the coordinates of the positives by looking for intensity peaks. The software will generate a simple text file containing the plate number and well address of the positive clones. The plate number should then be corrected for the order in which they were gridded on the membrane. The Xdigitise tool (H. Griffith, unpubl,; http://www.molgen.mpg.de/~xdigitise) is one example of such software.

*Acknowledgements.* The protocols described in this chapter are the results of several years of development performed first at the Genome Analysis Laboratory, ICRF (London), and then at the MPIMG (Berlin) by several teams of people who have improved many steps. These are Lisa Gellen, Holger Hümmerich, Susan Kirby, David Bancroft, Elmar Maier, Dean Nizetic, Mark Ross, and Günther Zehetner.

## References

Bancroft DR, O'Brien JK, Guerasimova A, Lehrach H (1997) Simplified handling of high-density genetic filters using rigid plastic laminates. Nucleic Acids Res 25:4160–4161

Cano RJ, Torres MJ, Klem RE, Palomares JC (1992) DNA hybridization assay using Attophos, a fluorescent substrate for alkaline phosphatase. Biotechniques 12:264–267

Höltke HJ, Ankenbauer W, Mühlegger K, Rein R, Sagner G, Seibl R, Walter T (1995) The Digoxigenin (DIG) system for non-radioactive labelling and detection of nucleic acids - an overview. Cell Mol Biol 41:883–905

Maas R (1983) An improved colony hybridization method with significantly increased sensitivity for detection of single genes. Plasmid 10:296–398

Maier E, Roest Crollius H, Lehrach H (1994) Hybridisation techniques on gridded high density DNA and in situ colony filters based on fluorescence detection. Nucleic Acids Res 22:3423–3424

Roest Crollius H, Jaillon O, Dasilva C, Ozouf-Costaz C, Fizames C, Fischer C, Bouneau L, Billault A, Quetier F, Saurin W, Bernot A, Weissenbach J (2000) Characterization and repeat analysis of the compact genome of the freshwater pufferfish *Tetraodon nigroviridis.* Genome Research 10:939–949

Scholler P, Heber S, Hoheisel JD (1998) Optimisation and automation of fluorescence-based DNA hybridization for high-throughput clone mapping. Electrophoresis 19:504–508

## ▨ Suppliers

- Robotic gridding machines
  BioRobotics Limited, 3-4 Bennell Court, Comberton,
  Cambridge, CB3 7DS, UK (e-mail: sales@biorobotics.co.uk,
  Tel.: +44-1223-264345, Fax: +44-1223-263933)
  Genetix, 63-69 Somerford Road, Christchurch, Dorset BH23
  3QA, UK (e-mail: sales@genetix.co.uk, Tel.: +44-1202-
  483900, Fax: +44-1202-480289)

- CCD camera
  Diagnostics Instruments Inc.
  6540 Burroughs
  Sterling Heights, Michigan 48314-2133, USA
  (Tel.: +1-810-731 6000, Fax: +1-810-731 6469)

- Pouch laminator
  GMP Ltd. D3 Telford Rd, Bicester, Oxon OX6 OTZ, UK
  (Tel.: +44-1869-323132, Fax: +44-1869-325 909)

- Blotting paper:
  Schleicher & Schuell, Inc. GmbH, P.O. Box 4, 37582 Dassel,
  Germany (e-mail: techserv@s-and-s.com, Tel.: +49-5561-
  7910, Fax: +49-5561-791536)

- Attophos
  JBL Scientific Inc., 277 Granada Dr., San Luis Obispo,
  California 93401, USA (e-mail: sales@jblsci.com,
  Tel.: +1-888-5448524, Fax: +1-805-5431531)

- Pronase, DIG-dUTP, anti-DIG antibody conjugated
  to alkaline phosphatase:
  Roche Molecular Biochemicals, Sandhoffer Strasse 116,
  68305 Mannheim, Germany
  (e-mail: mannheim.biocheminfo@roche.com,
  Tel.: +49-621-759 8540, Fax: +49-759-4083)

- Positively charged nylon membranes:
  Amersham Pharmacia Biotech AB, 751 84 Uppsala, Sweden
  (Tel.: +46-1816-5000, Fax: +46-1816-6458)

- Low fat milk
  Any local supermarket. Take a plain brand, with no fancy
  additives.

# Identification of Cell Targeting Ligands Using Random Peptide-Presenting Phage Libraries

TATIANA I. SAMOYLOVA and BRUCE F. SMITH

## ▓ Introduction

Phage display is a powerful technique for identifying peptides or proteins that bind to particular targets. The technique first described by Smith (1985), allows among other applications, rapid screening of random peptide libraries expressed on the surface of filamentous bacteriophage for ligands that can then be used for diagnostic and therapeutic needs.

Three filamentous phage, M13, fd and fl which are collectively named as the Ff phage, are the most commonly used expression vectors for phage display (reviewed in Webster 2001). The wild-type virion is about 6.5 nm in diameter and 930 nm in length. The Ff phage have 11 genes. Genes III, VI, VII, VIII, and IX encode capsid proteins of the phage particle. Two of them, gene III and gene VIII, have been most commonly used in phage display. pVIII, the product of gene VIII, is a major coat protein, and present in approximately 2700 molecules which encapsulate the phage. Gene III encodes a minor coat protein at the terminus of the filamentous phage particle. pIII is present in 5 copies, is required for infection of *E. coli*, and binds to the end of the F pili of the bacteria (Marvin, 1998).

Random peptide-presenting libraries have been constructed by the insertion of a DNA fragment, fixed in length but with random codons, in a phage surface protein gene, either III or VIII.

T.I. Samoylova, B.F. Smith (e-mail: smithbf@mail.auburn.edu,
Tel.: +1-334-8445587, Fax: +1-334-8445850)
Scott-Ritchey Research Center, College of Veterinary Medicine, Auburn
University, Auburn, Alabama 36849, USA

Springer Lab Manual
R.C. Bird, B.F. Smith (Eds.) Genetic
Library Construction and Screening
© Springer-Verlag Berlin Heidelberg 2002

The insertion of random oligonucleotide sequences results in a fusion protein on the phage surface. Thus, the expressed fusion peptides are directly linked to the genes encoding them which allows the identification of amino acid sequences. Libraries have been generated and used, displaying peptides ranging in size from 5 (Rickles et al. 1995) to 40 (McConnell et al. 1996) random amino acid residues.

Peptides exposed on the phage surface are available to act as a ligand, enzyme, enzyme inhibitor, immunogen or otherwise (reviewed with emphasis on different applications in Cortese et al. 1996; Daniels and Lane 1996; McGregor 1996; Katz 1997; Lowman 1997; Zwick, 1998; Rodi and Makowski 1999; Hoess 2001; Irving et al. 2001). To select specific peptides, a screening process in which binding clones are separated from nonbinding clones is accomplished by affinity selection in a method called biopanning. Efficient and simple protocols have been described to screen peptides binding to diverse targets such as purified proteins, antibodies, serum samples, cell surface molecules in vitro and in vivo, etc. Experiments in murine muscle cell cultures identified two 20-mer peptide sequences from a random peptide library which had enhanced muscle binding (Barry et al. 1996). Another group was able to identify peptides specific for renal and brain endothelium by screening a phage presentation library in vivo (Pasqualini and Ruoslahti 1996). The approach based on in vivo screening produced peptides capable of mediating selective localization of phage to tumor vessels with potential use for the delivery of therapeutics into selected tissue (Pasqualini et al. 1997; Arap et al. 1998; Koivunen et al. 1999).

This chapter will concentrate on using random peptide-presenting phage libraries for identification of cell targeting ligands. The selection protocol is simple to accomplish and does not require any special equipment. It takes no more than three days to complete one round of selection. The method needs no prior information about the target cells and has the advantage that it is directly applicable to target molecules in their native environment on cell surfaces. A large panel of cell binding peptides may be isolated to increase the likelihood that the most specific and highest affinity peptides will be identified. The binding found could readily be exploited to target appropriate vectors, constructs or pharmaceuticals or/and to isolate and characterize novel cell-specific receptors.

The majority of the peptide libraries available consist of about $10^7$–$10^9$ independent clones, which include all the possible 5–7-mer sequences. Libraries of longer peptides are incomplete in this respect but have been used successfully for ligand identification, confirming that the region which defines the binding specificity of a peptide contains a few amino acids. Theoretically, linear peptides can assume a number of conformations and, due to their flexibility, have the potential to bind to different targets. Therefore, libraries of cyclic (constrained) peptides are often considered to be richer sources of specifically binding molecules than linear peptides. However, it has been shown that short linear peptides can demonstrate a high-affinity binding to target protein as well (see reviews mentioned above). When choosing a library, one might consider multivalent phage display vectors as offer such features as high specificity and affinity (Petrenko et al. 1996, Ivanenkov et al. 1999; Romanov et al. 2001).

Random peptide libraries which have been constructed and used are listed in an issue of Chemical Review (Smith and Petrenko 1997). Many libraries are available from commercial and non-commercial sources. New England Biolabs markets random peptide libraries as kits: Ph.D.-7 Peptide 7-mer Library Kit (library of linear peptides), Ph.D.-12 Peptide 12-mer Library Kit (library of linear peptides), Ph.D.-C7C Disulfide Constrained Peptide Library Kit (library of cyclic peptides). Some non-commercial sources of random peptide libraries, including pIII/6-mer, pIII/15-mer, pVIII-4/15-mer libraries, etc. (George Smith, University of Missouri, USA), can be found on the Internet (http://www.biosci.missouri.edu/SmithGP).

This chapter will focus on protocols using New England Biolabs pIII phage display kits. Methods for working with peptide libraries displayed by the major coat proteins pVIII are described by Bonnycastle et al. (2001). Modifications to the protocols proposed here may be needed to achieve the optimal screening conditions for each particular application.

## ▪ Outline

To identify peptide ligands which bind to specific cells, two protocols are provided in this chapter. The first describes the steps involved in in vitro phage display peptide library screening on cultured cells. Phage selection in vitro is performed as described by Barry et al. (1996) with modifications. The second protocol may be used for tissue/organ-specific ligand selection in vivo in mice (Samoylova and Smith 1999), and with scale modifications in dogs. In addition, accompanying protocols for phage titering, plaque amplification and isolation of phage DNA, and confirming tissue specificity in vivo are provided.

Figure 1 shows the overall relationship of the steps involved in in vitro selection of cell-specific ligands. The procedure starts

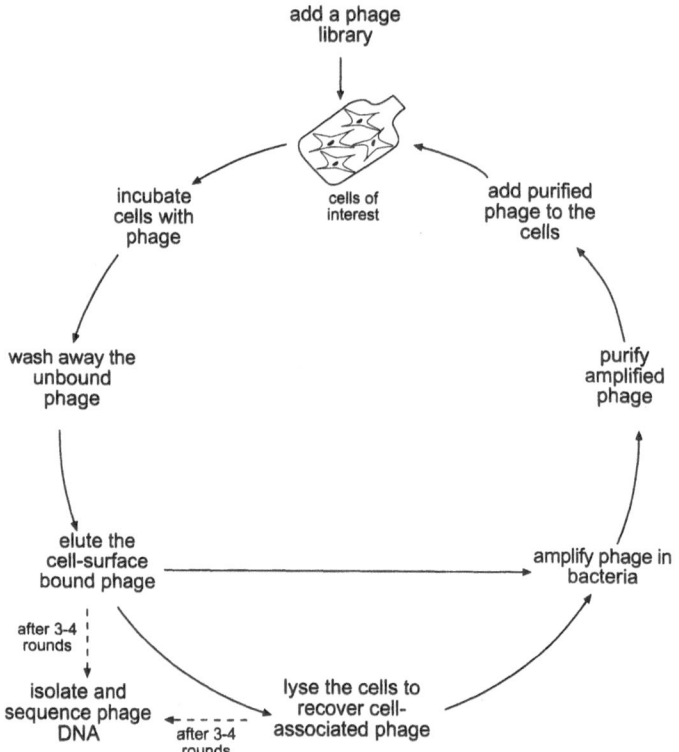

**Fig. 1.** Scheme for in vitro selection of cell-specific ligands using random peptide-presenting phage libraries. A phage library is added to a flask with cultured target cells and incubated for a fixed time. During the incubation, phage that display peptides specific to the cell-surface molecules are bound to the target. Unbound phage are then washed away. The membrane-binding phage fraction is eluted from the cell surface. Subsequently, the cells are lysed and the cell-associated phage fraction is recovered. Phage obtained are amplified in bacteria, purified, and added to target cells for each following round of selection. After three to four rounds, individual phage clones are sequenced

with incubation of the cells with a primary phage library in the presence of a blocking agent. One hundred equivalents of the original library are added to the cells. During this incubation, phage particles that display ligands specific to the cell surface molecules are bound to the target while nonspecific sites on the cells are blocked with bovine serum albumin (BSA). Unbound phage are then washed away very thoroughly with washing buffer alone or supplemented with detergent (e.g., Tween-20). Membrane-associated phage, which represent cell-binding peptides, are eluted from the cell surface by intense pH changes (e.g., glycine-HCl, pH 2.2; ethanolamine, pH 12) or other means, including 6 M urea or 50 mM dithiothreitol elution (Smith and Scott 1993; Sparks et al. 1996). Subsequently, the cells are lysed with lysis buffer and the cell-associated fraction, which represents internalized phage clones, is recovered. Both fractions obtained are amplified in bacteria (phage are grown on a complementing host) and purified by polyethylene glycol precipitation. Thus, in this method there is a two-phase process for recovering bound phage: cell-surface phage elution followed by cell lysis. The amplified eluate typically has a titer of $10^{10}$–$10^{12}$ pfu/ml which is a high enough phage number to accomplish the next round of selection.

Usually, a single round of selection is not sufficient to purify binding clones from a complex library. The repetition of the selection process allows enrichment of specific binding phage clones. Following three to four rounds of screening on cells and amplification in bacteria, individual phage clones are sequenced. The enrichment of binding phage between rounds of selection can be monitored by titering and sequencing of the phage DNA from each round. Selectivity of each phage clone can be confirmed in an additional experiment by the comparison of its binding to a panel of different cell lines including the cells they were selected against (Barry et al. 1996).

The in vivo ligand identification procedure, which is shown schematically in Fig. 2, allows peptides to be obtained that home selectively to different tissues and organs. The phage are administered to a mouse via tail vein injection and left in the circulation for a certain period of time depending on the targeted tissue and purpose of the experiment. After that time period, the mouse is sacrificed, perfused and the organs and tissues of interest are removed. To recover the bound phage, aliquots of each tissue are taken, washed and then homogenized in lysis buffer. The homog-

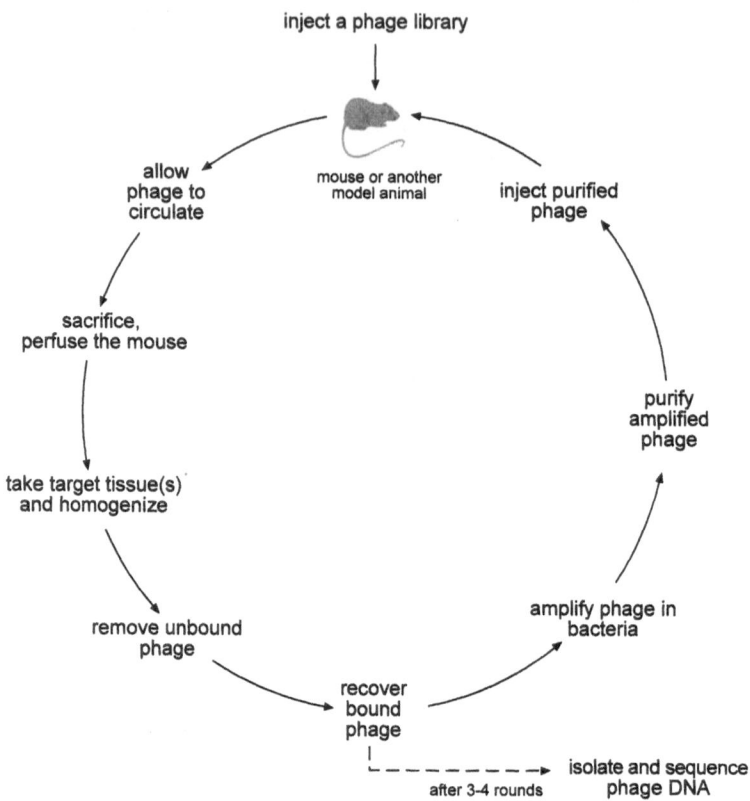

**Fig. 2.** Scheme for in vivo selection of cell-specific ligands using random peptide-presenting phage libraries. A phage library is administered to a mouse and left in the circulation for a certain period of time. After that, the mouse is killed, perfused and bound phage are recovered from the organs and tissues of interest. Phage obtained are amplified in bacteria, purified, and reinjected into mice for each following round of in vivo screening. After three to four rounds of screening, individual phage clones are isolated, and phage DNA is sequenced

enized tissue containing phage is used for phage titer determination and phage amplification in bacteria. Amplified and purified phage are reinjected into mice for each subsequent round of in vivo screening. After three to four rounds of screening, individual phage clones are isolated, and phage DNA is sequenced. The procedure described leads to the isolation of sequences with binding affinity to specific tissues and organs.

Candidate peptide ligands generated by in vivo selection may be evaluated for the ability to accomplish targeted binding

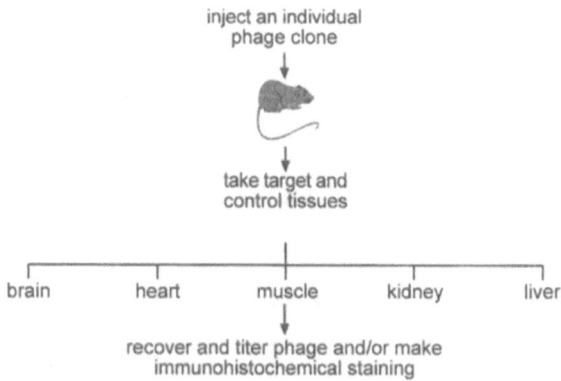

inject an individual
phage clone

take target and
control tissues

brain        heart        muscle        kidney        liver

recover and titer phage and/or make
immunohistochemical staining

**Fig. 3.** Scheme for confirming tissue specificity of selected phage in vivo. A pure phage clone is injected intravenously into mice. After 15 min to 1 h the mice are killed, perfused, and organs and tissues of interest are removed. Aliquots of each tissue are taken to determine the titer of phage recovered. Additionally, phage specific to target cells can be demonstrated by immuno-histochemical localization of the phage in tissue sections

(Fig. 3). To confirm tissue specificity, the clones carrying selected peptides are isolated and purified as individual clones. A pure clone is injected into mice and subsequently rescued from individual organs and tissues as in the in vivo selection procedure. Aliquots of tissues are taken to determine the titer of phage, without amplification. Wild-type phage lacking recombinant peptides are used as a control for the involvement of other phage associated proteins in cell binding. Specific binding of the selected clone is estimated on the basis of phage distribution between organs and tissues in control (wild-type phage) and test (pure phage clone) experiments. Additionally, phage targeting specific cells can be demonstrated by immunohistochemical localization of the phage in tissue sections. Following intravenous injection, tissue samples are obtained from mice and stained with an antibody against M 13 (Pasqualini et al. 2001).

## ▨ Materials

**Blocking/washing buffer** (used in in vitro phage selection protocol):
DMEM (Dulbecco's modification of Eagles medium) or another medium which is appropriate for the chosen cell line
0.1 % BSA
0.1 % Tween-20
Filter sterilize, store at 4 °C (better freshly prepared)

**Elution buffer** (for elution of cell surface binding phage in in vitro phage selection protocol):
0.1 M HCl, pH 2.2 (by glycine)

Prepare aqueous solution. Autoclave, store at room temperature.

**Lysis buffer** (for cell lysis in in vitro/in vivo phage selection protocols):
30 mM Tris-HCl, pH 8.0
1 mM EDTA
1 % Trotpm X-100

Prepare aqueous solution. Autoclave, store at room temperature.

**LB medium** (for liquid bacterial culture):
25 capsules of LB-Medium (Bio 101, Inc., cat # 3002–031) per 1 liter of purified water.

Autoclave at 121 °C for 15 min. Mix well and store at room temperature.

**2 × M9 salts** (for preparation of Minimal plates, see below):
12 g $Na_2HPO_4$
6 g $KH_2PO_4$
1 g NaCl
2 g $NH_4Cl$

Prepare 1 liter of aqueous solution. Autoclave, mix well. Store at room temperature.

**Minimal plates** (to maintain the bacterial culture):
100 ml 2 × M9 salts
4 ml 20 % glucose
400 μl 1 M MgSO$_4$
20 μl 1 M CaCl$_2$
200 μl thiamine (10 mg/ml)
100 ml 3 % agarose

Autoclave all components (with the exception of glucose and thiamine) and cool to less than 70 °C. Filter sterilize glucose and thiamine. Combine and mix the components (with the exception of agarose). Add the mixture to the agarose. Mix well and pour plates.

**Phage precipitation buffer** (used in phage purification procedure):
20 % (w/v) polyethylene glycol 8000 (PEG-8000)
2.5 M NaCl

Prepare aqueous solution. Autoclave, store at room temperature.

**TBS** (Tris Buffered Saline, used in phage purification procedure):
50 mM Tris-HCl, pH 7.5
150 mM NaCl

Prepare aqueous solution. Autoclave, store at room temperature.

**XGAL-IPTG** (for preparation of LB-Xgal-IPTG plates):
200 mg Xgal (5-Bromo-4-Chloro-3-Indolyl-β-D-Galactopyranoside)
250 mg IPTG (Isopropyl-β-D-Thiogalactoside)
5 ml dimethyl formamide

Mix well, store at –20 °C in the dark.

**LB plates** (for phage titering):
40 capsules of LB-Agar Medium (Bio 101, Inc., cat # 3002–231) per 1 liter of purified water.

Autoclave at 121 °C for 15 min. Cool to 50 °C. Mix well and pour plates. Store at 4 °C.

**LB-Xgal-IPTG plates** (to control the library phage contamination with wild-type phage):
Prepare 1 liter of LB-Agar Medium as for LB plates (see above). Autoclave, cool to 50 °C, add 1 ml Xgal-IPTG. Mix well and pour plates. Store at 4 °C in the dark.
Top agarose (for phage titering):
LB medium (see above)
1 g $MgCl_2 \times 6H_2O$
7 g agarose

Autoclave at 121 °C for 15 min. Dispense into 30-ml aliquots and store at room temperature. Melt in microwave as needed.

**Iodide buffer** (for phage DNA isolation):
10 mM Tris-HCl, pH 8.0
1 mM EDTA
4 M NaI

Prepare aqueous solution. Store at room temperature in the dark.

**TE buffer** (used to dissolve phage DNA):
10 mM Tris-HCl, pH 8.0
1 mM EDTA

Prepare aqueous solution. Autoclave, store at room temperature. One can use purified water (Tissue Culture Water, Sigma, cat# W-3500) as well.

## ▨ Procedure

### Protocol 1. Affinity selection of cell binding phage in vitro (on cells)

1. Grow cells of interest in Corning (cat # 430168)/Falcon (cat # 3081) 25-cm² flasks using appropriate media for chosen cell lines.

**Day 1 (phage selection procedure on cells)**

2. Incubate cells in 2 ml of serum-free medium (DMEM or another appropriate for chosen cell line) for 1 h before application of the phage.

3. Discard the serum-free medium. Add 100 equivalents of original phage library to the flask in 2 ml of blocking buffer for 1 h incubation at room temperature.

4. Carefully remove the medium with nonbound phage from the flask by using a pipette. Wash cells six times with 4 ml of cold washing buffer for 5 min. Each time, remove the washing buffer very carefully. Do not leave any remainder in the flask.

5. Incubate cells for 10 min on ice with 2 ml of elution buffer to elute cell-surface bound phage. Pipette eluate into a sterile tube and neutralize with 400 µl 1 M Tris-HCl, pH 8.0. Save this fraction as cell-surface phage and store at 4 °C until the next day for amplification.

6. Wash cells two times as in step 4. Remove the washing buffer.

7. Add 1 ml of lysis buffer for 1 h on ice to recover the cell-associated phage fraction. Scrape the cell monolayer from the bottom of the flask with a sterile spatula. Harvest the cell debris in lysis buffer by pipetting and store at 4 °C until the next day for amplification.

8. Inoculate 20 ml of LB medium with a single colony of bacteria (an appropriate *E. coli* host strain, for example, *E. coli* ER2738 supplied by New England Biolabs) in, for example, 50 ml screw cap conical Sarstedt tubes (cat # 62.547.004), one for each phage fraction (cell-surface and cell-associated, see Day 1 above), and incubate at 37 °C with shaking until early-log phase. **Day 2 (phage amplification in bacteria)**

9. Add the phage eluate to the bacterial culture (each fraction to a separate tube) and incubate at 37 °C with vigorous shaking for 4.5 h.

10. Transfer the culture to sterile centrifuge tubes (e.g. Nalgene, cat # 3115–0050) and pellet any bacterial cells by centrifugation at 14,000 $g$ (10,000 rpm, SA-600 rotor in a Sorvall RC5C centrifuge) for 15 min at 4 °C. Phage particles remain in the supernatant.

11. Pipette the supernatant to a fresh centrifuge tube. Precipitate the phage by adding one sixth volume of precipitation buffer. Mix well and leave at 4 °C overnight.

**Day 3 (phage purification)**

12. Pellet the precipitated phage by centrifugation at 14,000 $g$ (10,000 rpm, SA-600 rotor in a Sorvall RC5C centrifuge) for 20 min at 4 °C. Decant supernatant and invert the tube over a paper towel to drain off the residual liquid.

13. Resuspend the pellet in 1 ml TBS by pipetting up and down with a micropipette. If desired, transfer the phage suspension to a microcentrifuge tube and spin for 5 min at 4 °C to eliminate any residual bacterial cells.

14. Transfer the supernatant to a fresh microcentrifuge tube and re-precipitate by adding one sixth volume of precipitation buffer and follow on with incubation on ice for 1 h.

15. Microcentrifuge for 10 min at 4 °C to pellet the precipitated phage. Carefully remove any supernatant with a micropipette.

16. Suspend the pellet in 100 μl TBS, 0.02 % $NaN_3$. Microcentrifuge the resuspended phage for 10 min to pellet any insoluble material. Transfer the supernatant (the amplified phage eluate) to a fresh tube and store at 4 °C. For long term storage add 50 % of sterile glycerol and store at –20 °C.

17. To quantitate the phage, titer the amplified eluate as described below (see Protocol 3, Phage titering).

**Day 4 (next round of selection)**

18. Calculate the input volume of amplified phage for the next round of selection based on the phage titer determined. Use the same number of phage as for the first round of selection.

19. Enrichment of phage with affinity for specific cells is then done by successive rounds of selection. Carry out the next round of selection by repeating steps 2–17.

20. After 3–4 rounds, isolate DNA of individual phage clones (see Protocol 4, Plaque amplification and isolation of phage DNA) and sequence phage DNA.

## Protocol 2. Affinity selection of tissue-specific phage in vivo

1.  A mouse is put into a restraining cage (Lab Products, Inc.) and injected intravenously via a tail vein with a pool of phage containing 100 equivalents of original phage library diluted in 200 μl PBS. <u>Note:</u> Some strains of mice(for example, nude mice) are sensitive to bacterial endotoxins, so phage should be purified prior to injection. Several methods for endotoxin removal from phage preparations have been compared in our laboratory. Triton X-114 (Sigma, cat # 9036–19–5) phase separation has been found to be the optimal one.  

    *Day 1 (phage selection procedure in vivo)*

2.  After 15 min to 1 h the mice are euthanized (preferably by $CO_2$ narcosis), and perfused via the left ventricle with approximately 50 ml ice-cold PBS/heparin, 1000 units heparin per 1 liter PBS. **Attention:** this step is critical for successful selection. Accurate perfusion allows a substantial reduction in the background phage left in the circulation. To standardize the perfusion one can use a syringe pump at the rate of 50 ml in 15 min.

3.  Organs and tissues of interest are removed and immediately placed in ice-cold PBS. 100 mg aliquots of each tissue are finely diced with a scalpel and homogenized in 1 ml of PBS using a manual homogenizer. **Note:** If needed, retain aliquots (10 μl) of homogenized tissue to determine the titer of phage recovered. Protease inhibitors such as 1 mM PMSF, 20 μg/ml aprotinin and 1 μg/ml leupeptin may be used at his step.

4.  Microcentrifuge the homogenates for 10 min to collect the bound phage. Remove the supernatant containing unbound phage. Resuspend the pellet in PBS and repeat step 5 again to minimize background phage.

5.  Resuspend the pellet in lysis buffer and keep in a refrigerator overnight. Use this tissue suspension for phage amplification and the next round of selection.

6.  Inoculate 20 ml of LB with a single colony of bacteria in a sterile tube, for example, 50 ml screw cap conical Sarstedt tubes (cat # 62.547.004), one for each tissue, and incubate at 37 °C with shaking until early-log phase.  

    *Day 2 (phage amplification in bacteria)*

7. Add the tissue containing bound phage (step 5) to the bacterial culture and incubate at 37 °C with vigorous shaking for 4.5 h.

8. Follow the amplification procedure for phage selected in vitro (Protocol 1, steps 10, 11).

**Day 3 (phage purification)**

9. Purify phage as described in Protocol 1, steps 12–16.

10. To quantitate phage, titer the amplified phage as described in Protocol 3, Phage titering.

**Day 4 (next round of selection)**

11. Count plaques and determine phage titer. Use this value for the next round of selection to calculate an input volume of amplified phage for the injection. Use the same phage number as for the first round of selection.

12. Carry out the next round of in vivo selection: repeat steps 1–3, Protocol 2. <u>Attention</u>: if the selection process is carried out for more than one kind of tissue, the corresponding number of mice should be used in the second and each following round of selection, i.e., one mouse per targeted tissue.

13. After 3–4 rounds, isolate DNA of individual phage clones (Protocol 4, Plaque amplification and isolation of phage DNA) and sequence phage DNA.

**Protocol 3. Phage titering**

1. Grow a bacterial culture (one needs 200 μl of bacterial culture per sample to be titered) at 37 °C with shaking until an $OD_{600}$ of about 0.5 is reached, which takes about 6–10 h (strain-dependent).

2. Set up tenfold serial dilutions of phage in LB medium depending on suggested phage concentration (titer) in the sample. The titer may vary between $10^2$–$10^6$/g wet tissue for unamplified phage in the tissue samples obtained after in vivo selection, and $10^8$–$10^{12}$/ml for amplified phage. Keep diluted phage on ice until use.

3. Prewarm LB plates at 37 °C, and keep them in the incubator until ready for use. The number of plates should be equal to the number of dilutions one plans to plate (see step 2 above).

**Note:** If the library phage carry the lacZ reporter gene, use LB-Xgal-IPTG plates to monitor the library phage contamination with wild-type phage. The library phage form blue plaques when plated on media containing Xgal and IPTG. Wild-type phage form white plaques on that media.

4. Melt top agarose in microwave and dispense 3 ml into sterile tubes, one per phage dilution. Transfer tubes to a 55 °C bath, allow the temperature to equilibrate, and keep until use. Make sure that the temperature does not fall below 55 °C or the agarose will set.

5. Dispense 200 μl bacterial culture into sterile microfuge tubes, one for each phage dilution.

6. Take 10 μl of each dilution and add to each of the tubes with bacterial culture that were prepared in step 5, vortex quickly, and incubate at room temperature for several minutes.

7. One at a time, transfer the infected bacteria, that were prepared in step 6, to a tube containing 3 ml prewarmed to 55 °C top agarose, mix briefly with a vortex, and pour over the top of a prewarmed LB plate very quickly. Tilt the plate to spread top agarose along the entire surface.

8. Allow the top agarose to harden at room temperature for 5 min, invert plates and leave overnight in a 37 °C incubator.

9. Count the number of plaques (transparent foci, 0.5–1 mm in diameter) on plates having about 100 plaques. Determine the titer of the phage by multiplying the number of plaques by the dilution factor for that plate and the volume plated. For example, if one took 10 μl of phage suspension for titering and the number of plaques on the plate with $10^{-10}$ dilution is equal to 47, the titer of the phage per ml will be equal to $4.7 \times 10^{13}$/ml ($47 \times 10^{10}/10$ μl$=4.7 \times 10^{13}$/ml).

**Protocol 4. Plaque amplification and isolation of phage DNA**

1. Grow a bacterial culture overnight at 37 °C with vigorous shaking.

2. Dilute the culture 1:100 in LB medium and dispense 1 ml diluted culture into 6 ml Falcon culture tubes (cat # 2063). Prepare one tube per phage clone to be amplified.

3. Use a sterile micropipette tip. Stab random plaques on the plate and drop into a tubes containing bacterial culture. **Attention:** pick plaques from plates having no more than 100 plaques to be sure that each plaque contains an individual phage clone. Incubate tubes at 37 °C with shaking for 4.5–5 h.

4. Transfer cultures to microcentrifuge tubes, and centrifuge for 30 s to pellet bacteria. Transfer the supernatant to a fresh tube and repeat the step.

5. Using a pipette, transfer the upper 700 µl of the phage-containing supernatant to a fresh microfuge tube. Save the rest of the supernatant (approximately 300 µl) in case you need to propagate this particular phage clone to confirm its tissue specificity in vivo (see Protocol 5). The phage may be stored at 4 °C for several weeks.

6. To precipitate the phage, add 200 µl precipitation buffer. Invert to mix, and let stand at room temperature for 10 min.

7. Centrifuge for 10 min to pellet phage particles. Discard supernatant, re-spin briefly, and carefully pipette away any remaining supernatant.

8. Suspend pellet thoroughly in 100 µl of iodide buffer. Add 250 µl 95 % ethanol, mix by inversion, and incubate for 10 min at room temperature. A short incubation at room temperature will preferentially precipitate single-stranded phage DNA, leaving most phage protein in solution.

9. To harvest phage DNA, spin for 10 min. Discard supernatant. Note: keep in mind that the DNA pellet is often invisible at this step.

10. Wash pellet twice in 500 µl 70 % ethanol. Spin for 10 min in a microfuge each time. Discard the ethanol and invert to drain.

11. Vacuum dry the pellet for 5–10 min or until dry. Suspend pellet in 20 μl TE buffer or sterile water. Determine DNA concentration by spectroscopy. Store at –20 °C. Phage DNA prepared in this manner is suitable for automated or manual sequencing.

### Protocol 5. Confirming tissue specificity in vivo

1. Grow a bacterial culture overnight at 37 °C with vigorous shaking.

2. Add phage left in step 5 in Protocol 4 (Plaque amplification and isolation of phage DNA) to the bacterial culture and incubate at 37 °C with vigorous shaking for 4.5 h.

3. Purify individual phage clone following the purification procedure for phage selected in vitro (Protocol 1).

4. Mice are injected intravenously via a tail vein with $10^{10}$ pfu of individual phage clone diluted in 200 μl PBS. Phage dose may vary depending on target tissue.

5. Repeat step 2 from Protocol 2.

6. To recover the bound phage, 100 mg aliquots of each tissue are homogenized in 1 ml of lysis buffer using a manual homogenizer. Take an aliquot (10 μl) of homogenized tissue to determine the titer of phage recovered.

7. Use the absolute number (pfu) of control and test phage per gram of wet tissue recovered from targeted and control tissues to confirm tissue specificity of selected phage clone.

## ▪ Results

To determine the potential cell-specific peptide ligands in each screening experiment, phage DNA from about 40 clones is sequenced in our laboratory. If no similar or identical clones are found, rescreening of the current library under modified conditions or application of other libraries, such as those containing constrained peptides, peptides of various length or multivalent libraries, is recommended (see optimization procedure below).

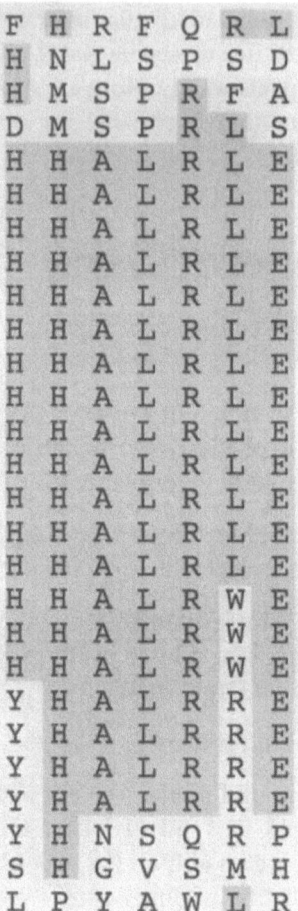

**Fig. 4.** Peptide sequences from in vivo phage display screening in murine skeletal muscle which are related to the consensus sequence $^H/_Y$ H A L R $^L/_W/_R$ E

Another result one might see would be the identification of several groups of identical clones among the sequenced phage clones with no similarity between the groups. In this case, one needs to test all the dominant clones for tissue specificity. If several related but not identical groups of clones are discovered a conclusion about the consensus sequence responsible for binding to the target can be made. An example of this occurred when 38 phage clones were sequenced after three rounds of selection in vivo in murine skeletal muscle (Samoylova and Smith, unpubl. data). The results revealed a number of similar, but not identical, sets of sequences that represented more than 80 % of all clones (Fig. 4). Comparative analysis of these sequences revealed that the most common sequence was HHALRLE, present in 13 clones with the sequence HHALRWE present in three clones and the sequence YHALRRE present in four clones. These sequences may represent an optimal consensus sequence responsible for muscle binding.

## Procedure optimization

A simple and effective capture technique for the selection of specific ligands from a library of peptides expressed on the surface of bacteriophage is described in this chapter. However, the nature, variety and density of potential target(s) are unique for each cell type and in most cases unknown. Under these circumstances, approaches for procedure optimization are proposed, rather than a troubleshooting guide for identification and solving of potential problems.

To optimize peptide-target interactions when using the in vitro protocol, one can modify screening conditions such as input concentration of phage, duration and temperature of incubation/elution steps, elution buffer used, number and temperature of washes, and concentration of detergent in the blocking/washing buffers. Recommended temperatures to try at each step of selection are 4 °C, room temperature and 37 °C. The concentration of detergent may vary from 0.1 to 0.5 % in blocking and washing buffers. Output titers in the eluates may be used for quantitative characterization of phage binding activity. An optimized method for cell-based phage display biopanning was described by Watters et al. (1997) using a model system of fixed cells. To increase the spcificity of binding, a depletion step may be performed prior to screening on target cells.

When choosing a selection protocol, one has to keep in mind that both protocols, in vitro and in vivo, have some advantages and disadvantages. While the selection on cultured cells has led to the identification of cell-specific peptides, there are some important considerations to make before using the sequences obtained in vitro for therapeutic applications. First, the search for the isolation of specific ligands is complicated by the tendency for a number of cultured cell lines to express fetal or embryonic genes rather than those of mature cells. In contrast, the in vivo ligand screening procedure leads to the identification of ligands for mature cells. Second, there are several physiologic barriers to in vivo screening, not encountered when using cell cultures, which may affect ligand-receptor recognition. For example, when screening for skeletal muscle, some phage clones may be bound in blood and/or by endothelial cells before reaching the myofibers. Only myofiber-specific phage clones which overcome these barriers can be selected. So, one can expect that these clones may be used successfully for gene/drug delivery to muscle when injected intravenously. Another advantage of in vivo selection is the possibility to screen against diseased tissues in animal models while such cell lines are often not available for cell culture. Finally, negative selection by untargeted tissues occurs in the course of in vivo screening that decreases background phage in the targeted tissue. On the other hand, one can use human cell lines for ligand selection in vitro, while the in vivo procedure requires the euthanasia of an animal and cannot be applied to humans. To solve these problems, in vitro and in vivo approaches may be combined or different species may be used in in vivo screening in order to find interspecies cell-specific ligands which are applicable for humans.

Development of inter-species ligands may be possible using multiple species, and could be enhanced by alternating animal species in the selection scheme. To adapt the in vivo protocol for screening binding peptides against particular tissues in different animal models, the following parameters may be taken into consideration: input phage, time in blood circulation, and perfusion conditions. A protocol for phage display library selection in dogs has been developed in our laboratory. While the procedure has led to the isolation of phage clones which bind preferentially to the tissues of interest, the procedure itself is difficult to perform due to the size of the animal and high number of phage required. The mass of tissue requires $10^{14}$–$10^{15}$ pfu of phage ($10^3$–$10^4$ equivalents of original phage library), and large samples of tissue combined from different sites. Finally, the size and circulatory volume of the dog make complete perfusion, and therefore removal of unbound phage, extremely difficult to perform in dogs. Therefore we have developed an assay using an acoustic wave sensor to examine the interspecies properties of phage display derived peptide ligands (Samoilov et al. 2002). The method allows detec-

tion of ligand-receptor interactions in tissue homogenates, open-
ing the possibility to estimate the apparent affinity of peptides
selected in one species to another, including humans.

## References

Arap W, Pasqualini R, Ruoslahti E (1998) Cancer treatment by targeted drug
    delivery to tumor vasculature in a mouse model. Science 279:377–380
Barry MA, Dower WJ, Johnston SA (1996) Toward cell-targeting gene therapy
    vectors: selection of cell-binding peptides from random peptide-present-
    ing phage libraries. Nat Med 2:299–305
Bonnycastle LC, Menendez A, Scott JK (2001) 15. General phage methods. In:
    Barbas III CF, Burton DR, Scott JK, Silverman GJ (eds) Phage display. A lab-
    oratory manual. Cold Spring Harbor Laboratory Press, Cold Spring Har-
    bor, New York
Cortese R, Monaci P, Luzzago A, Santini C, Bartoli F, Cortese I, Fortugno P, Gal-
    fre G, Nicosia A, Felici F (1996) Selection of biologically active peptides by
    phage display of random peptide libraries. Curr Opin Biotech 7:616–621
Daniels DA, Lane DP (1996) Phage peptide libraries. Methods 9:494–507
Hoess RH (2001) Protein design and phage display. Chem Rev 101:3205–3218
Irving MB, Pan O, Scott JK (2001) Random-peptide libraries and antigen-frag-
    ment libraries for epitope mapping and the development of vaccines and
    diagnostics. Curr Opin Chem Biol 2001 5:314–324
Ivanenkov VV, Felici F, Menon AG (1999) Targeted delivery of multivalent
    phage display vectors into mammalian cells. Biochim Biophys Acta
    1448:463–472
Katz BA (1997) Structural and mechanistic determinants of affinity and speci-
    ficity of ligands discovered or engineered by phage display. Annu Rev Bio-
    phys Biomol Struct 26:27–45
Koivunen E, Arap W, Rajotte D, Lahdenranta J, Pasqualini R (1999) Identifica-
    tion of receptor ligands with phage display peptide ibraries. J Nucl Med
    40:883–888
Lowman HB (1997) Bacteriophage display and discovery of peptide leads for
    drug development. Annu Rev Biophys Biomol Struct 26:401–424
Marvin DA (1998) Filamentous phage structure, infection and assembly. Curr
    Opin Struct Biol 8:150–158
McConnell SJ, Uveges AJ, Fowlkes DM, Spinella DG (1996) Construction and
    screening of M 13 phage libraries displaying long random peptides. Mol
    Diversity 1:165–176
McGregor D (1996) Selection of proteins and peptides from libraries displayed
    on filamentous bacteriophage. Mol Biotechnol 6:155–162
Pasqualini R, Ruoslahti E (1996) Organ targeting in vivo using phage display
    peptide libraries. Nature 380:364–366
Pasqualini R, Koivunen E, Ruoslahti E (1997) αv Integrins as receptors for
    tumor targeting by circulating ligands. Nat Biotech 15:542–546
Pasqualini R, Arap W, Rojotte D, ruoslahti E (2001) 22. In vivo selection of
    phage-display libraries. In: Barbas III CF, Burton DR, Scott JK, Silverman GJ

(eds) Phage display. A laboratory manual. Cold Spring Harbor Laboratory Press, Cold Spring Harbor, New York

Petrenko VA, Smith GP, Gong X, Quinn T (1996) A library of organic landscapes on filamentous phage. Protein Eng 9: 797–801

Rickles RJ, Botfield MC, Zhou XM, Henry PA, Brugge JS, Zoller MJ (1995) Phage display selection of ligand residues important for Src homology 3 domain binding specificity. Proc Natl Acad Sci USA 92:10909–10913

Rodi DJ, Makowski L (1999) Phage-display technology - finding a needle in a vast molecular haystack. Curr Opin Biotech 10:87–93

Romanov VI, Durand DB, Petrenko VA (2001) Phage display selection of peptides that affect prostate carcinoma cells attachment and invasion. Prostate 47:239–251

Samoylov AM, Samoylova TI, Hartell MG, Pathirana ST, Smith BF, Vodyanoy VJ (2002) Recognition of cell-specific binding of phage display derived peptides using an acoustic wave sensor. (Biomol Eng (in press)

Samoylova TI, Smith BF (1999) Elucidation of muscle binding peptides by in vivo phage display screening. Muscle Nerve 22:460–466

Smith GP (1985) Filamentous fusion phage: novel expression vectors that display cloned antigens on the virion surface. Science 228:1315–1317

Smith GP, Petrenko VA (1997) Phage display. Chem Rev 97:391–410

Smith GP, Scott JK (1993) Libraries of peptides and proteins displayed on filamentous phage. In: Wu R (ed) Recombinant DNA. Methods in enzymology, vol 217, part H. Academic Press, San Diego, pp 228–257

Sparks AB, Adey NB, Cwirla S, Kay BK (1996) Screening phage-displayed random peptide libraries. In: Kay BK, Winter J, McCafferty J (eds) Phage display of peptides and proteins. A laboratory manual. Academic Press, San Diego, pp 227–253

Watters JM, Telleman P, Junghans RP (1997) An optimized method for cell-based phage display panning. Immunotechnology 3:21–29

Webster R (2001) 1. Filamentous phage biology. In: In: Barbas III CF, Burton DR, Scott JK, Silverman GJ (eds) Phage display. A laboratory manual. Cold Spring Harbor Laboratory Press, Cold Spring Harbor, New York

Zwick MB, Shen J, Scott JK (1998) Phage-displayed peptide libraries. Curr Opin Biotechnol 9:427–436

## ▓ Suppliers

Becton Dickinson Labware (for Falcone disposable labware), 2 Oak Park, Bedford, Massachusetts 01730, USA (Tel.: #1-800-3432035, Fax: +1-781-2750043, www.bd.com/labware)

BIO 101, Inc., 1070 Joshua Way, Vista, California 92083, USA (Tel.: +1-800-4246101, Fax: +1-760-5980116, www.bio101.com)

Corning, Inc., Science Products Division, 45 Nagog Park, Acton,
Massachusetts 01720–9897, USA (Tel.: +1-800-4921110,
Fax: +1-978-6352478, www.scienceproducts.corning.com)

Lab Products, Inc., 742 Sussex Avenue, Seaford, Delaware 19973,
USA (Tel.: +1-800-5260469, Fax: +1-302-6284309)

Nalge Nunc International (for NALGENE Brand Products),
P.O. Box 20365, Rochester, New York 14602–0365, USA
(Tel.: +1-800-6254327, Fax: +1-800-NALGENE,
www.nalgenunc.com)

New England Biolabs, Inc., 32 Tozer Road, Beverly, Massachusetts 01915, USA (Tel.: +1-800-NEB-LABS, Fax: +1-978-9211350,
www.neb.com)

Sarstedt, Inc., P.O. Box 468, Newton, North Carolina 28658–0468,
USA (Tel.: +1-828-4654000, Fax: +1-828-4650718,
www.sarstedt.com)

Sigma, P.O. Box 14508, St. Louis, Missouri 63178, USA
(Tel.: +1-800-3253010, Fax: +1-800-3255052, www.sigma-
aldrich.com)

# Subject Index